The one hundred epic events of World War II stamp series as designed and printed by the Unicover Corporation of Cheyenne, Wyoming on behalf of the Republic of the Marshall Islands make it :

The best set of thematic stamps – EVER

as defined by Christopher B Yardley, PhD

A Cannava House Publication
Canberra, Australian Capital Territory
Cannava@iinet.net.au

Any part of the text may be reproduced by any process without written
permission. Enquiries should be addressed to the publisher.

National Library of Australia
Cataloguing-in-Publication entry
Yardley, Christopher B.
"The best thematic stamp set EVER"
ISBN : 978-0-6486671-5-5
EPub ISBN : 978-0-6486671-6-2

Keywords :
  i)   Second World War.
  ii)   World War II.
  iii)   Republic of Marshall Islands Postage stamps 1989 - 1995.
  iv)   History of human conflict 1939-1945.
  v)   Military History on postage stamps.

On the front cover are shown a 12-stamp sheet as published by Unicover, on
behalf of the Marshall Islands Postal Service of W1, the invasion of Poland in
1939 and also (at a different scan scale), a block of four stamps showing D-Day
activity of 1944.

Preface

The devil's in the detail, they say. And as far as the miniscule art form of stamp design is concerned, this couldn't be more true. The amount of contemplation and concentration involved in creating such a diminutive image is sizeable. Every pixel counts, every micro-millimetre needs to do its bit.

But of course, a stamp can only say so much. It's an impression, a small window to people and stories and celebrations, a cue to find out more.

By presenting (these) diverse subjects in summary and in detail, we hope to give you a fresh perspective. After all, our stamps may be small, but their scope and ambition are enormous (Editorial (2010), "The big picture", Royal Mail Year Book).

My project path to justifying "the best set of postage stamps EVER".

Introduction

I am a lifetime stamp collector. I was given responsibility for the "family" stamp collection at five and have respected that responsibility for eighty years – although, to be honest, it was costing more than I could justify from a pension and I have sold the larger part of the collection through auction by Status International of Sydney. At the time I was enjoying writing up my collection from a military historical perspective, and I still am.

I found that working with stamp images on the computer was as fulfilling as handling the actual paper object – particularly the most recent issues (being spoilt by the inclusion of QR codes into the design). I have looked at the world-wide issues commemorating The Great War, World War II and the other Wars of the Twentieth Century.

Could I hone down further and create a monograph on a single stamp issue? That is the challenge I have set myself. And why waste the opportunity to investigate what I consider to be "The best thematic set of postage stamps – EVER". I studied so many other sets of postage stamps but, to my mind, that best set was issued on behalf of the Republic of the Marshall Islands, by Unicover Corporation of Wyoming, during 1989 to 1995, to celebrate the fifty-year anniversaries of the One Hundred Epic Events of World War II.

I had discovered the set in studying the stamps representing the world conflict that was the Second World War.

I am aware that the Universal Postal Union (UPU) had in 2016 issued a revised Philatelic Code of Ethics for the use of its member countries when issuing and supplying postage stamps and other philatelic products.

According to the code, the issuing postal authorities shall not produce postage stamps or philatelic products that are intended to exploit customers. Features that abuse these criteria will be :

- An issue whose theme is a subject totally contrary to the culture of the issuing member country or territory, and which cannot be considered as contributing to "the dissemination of culture or to maintaining peace".
- An issue whose quantity far exceeds the acceptable limit for philatelic issues, where the number of annual issues is unrelated to the actual market capacity, whether for postal prepayment or for Stamp collection of the member country or territory concerned. (Philatelic Code of Ethics

for the use of UPU member countries, UPU recommendation C 13/2016).

I do not consider the set we are looking at breaches the UPU ethics. No one as far as this research goes has questioned the validity or integrity of the set. The Marshall Islands were very much involved in WW2. One might question an American bias, but this is not really apparent apart from a concentration upon General Douglas MacArthur, who was pursuing his own political aspirations, before the conflict ended. The issue and individual items from it are catalogued are freely available from reputable stamp dealers worldwide, and/or from Internet dealers such as eBay.

Don Sundman, of Mystic Stamp Company of New York sends out a daily sales-flyer, (to those who have asked him to) that always contains a potted history of the stamp issue he is writing about. This morning (24 February 2025) he discussed and shewed early evidence of the interest being showed by an "authority" about postage stamps they thought were "unnecessary or overpriced."

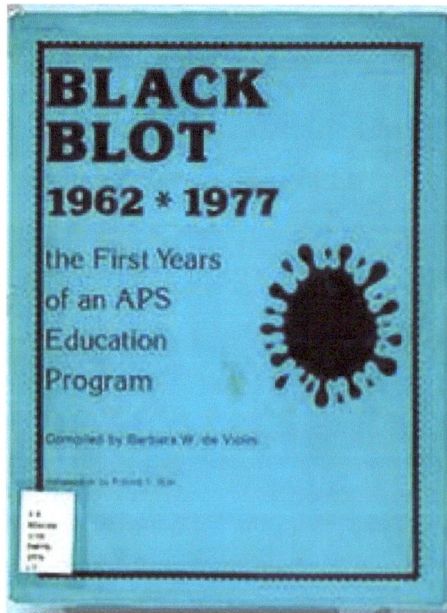

Created by the APS, containing stamps it thought were unnecessary or overpriced.

Don Sundman wrote :

Did you know the American Philatelic Society (APS) used to have a "black blot" list of stamps? It included all the stamps deemed to have "no postal necessity, excessive numbers of individual stamps, high values relative to the postal needs of the issuing country, or issues that included intentional errors or imperfs." Basically, it was a list of stamps the APS thought exploited the hobby by trying to intentionally appeal to stamp collectors.

The list originated in the 1960s and was updated with new issues through the early 1990s. But the practice was discontinued when it became apparent that most new US stamps should be black blotted.

When the first 50-stamp se-tenant was issued by the United States in 1976, it was almost immediately black blotted — most likely for excessive number of individual stamps. I liked the state flag stamps and thought collectors should decide what stamps to add to their collection. Personally, I thought the black blot idea was silly and was happy when the APS discontinued it. The new se-tenant format proved popular with collectors, despite APS warnings..."

I endorse Don Sundman's conclusions.

Republic of the Marshall Islands

157 stamp images that constitute a pictorial history of World War Two

Postage stamps are designed to convey messages that reverberate symbolically with broad swaths of the public, and their content has been employed as a window into how government, government bodies and /or their agents help members of the public understand the ideas / historical analysis represented therein.

Research and acquisition, paths taken, to get this far.

The question I was asking myself was "Is there a story here?".

1. Looking for an inside story of this stamp issue I sought guidance from the editor of *Themescene,* the journal of the *British Thematic Society* with whom I have conducted a long-term rapport over "when is a stamp not a (proper) stamp?" She advised me to seek 'advice' from 'the experts' at the *Royal Philatelic Society*. They have responded – their approach has been a search of articles in two of the respected stamp magazines; *Gibbons Monthly* and the *Linn's News*. Neither of these respected houses has featured the Marshall Islands set.

2. I have also made enquiries of the Philatelic Manager of the Smithsonian Institute in Washington. He confirmed that the Unicover Corporation of Wyoming, acted as Administrators of the Marshall Islands Postal Authority for the issue of interest

The Unicover Corporation business was founded by James A. Helzer (1946 – 2008) who started his career as an entrepreneur by selling old First Day Covers (a popular way to collect postage stamps) out of his parents' garage during junior high school. For almost half a century, Unicover built on that First Day Cover business as an innovative manufacturer and marketer of philatelic and numismatic commemoratives until acquired by the Mystic Stamp Company (of Camden, New York 13316)

James A Willms President and Chief Executive Officer of the Unicover Corporation of Cheyenne, Wyoming, after 33 years of service to the Company, retired in 2013. He was also President of Unicover World Trade Corporation, a wholly owned subsidiary responsible for Unicover's international business. He has advised me "*Unfortunately no history of the project itself was prepared. What I can tell you is that the concept itself and details like the location of the designers' name and commentary were the idea of then Uncover President, now deceased. I do not know of anyone else who can help you". "You might get some information from John Helzer, Jim's son, as I know he worked on some of the research and writing for the project. His email is : 'johnhelzer@gmail.com' Be sure to let him know that I referred you to him".*

After three requests the recipient of this e-address replied, "Wrong man". A further note to James Willms elicited a new e-address – *johnchelzer@gmail.* Great success – the correct / best person to move the project forward.

John Clark Helzer is the son of the founder and President of Unicover. Helzer senior has died but the son is most keen and was a member of the project team for the issue. As I had hoped Unicover, concluding the project, issued it its own history and review : The son has written *"I am thrilled to hear from you! Thank you so much for writing to me, and for your interest in my late father James A. Helzer's brainchild The One Hundred Epic Events of World War II in the postage stamp series issued by the Republic of the Marshall Islands (RMI), which was conceived of, designed, created and printed by my father's company Unicover Corporation (under the trademark of its flagship Fleetwood division) in the early years of its three-decade role as the Stamp Issuing Bureau of the Marshall Islands Postal Service Administration".*

John C Helzer has sent me a copy of the "coffee-table" style book published by Fleetwood, the first day (stamps) cover subsidiary of Unicover Corporation. This book provides answers to nearly all of the questions I have about the set of stamps of interest. John Helzer has also made himself available to answer any questions I might have. Those questions are largely answered in the book. Does the Fleetwood book make my research un-necessary? To a large part "yes". But my bringing it back to todays' audience makes it worthwhile.

3. The National Library of Australia have looked through their "Trove System" on my behalf to seek any references to the set within Australian newspapers and specific magazines within their interest. They recommended an approach to the Library of South Australia. This Library has been through its on-line resource and found a Gibbons Magazine reference from a 1977 issue on the subject of the postal systems in Micronesia – of interest but not of value to this story.

4. Several direct approaches to the Mystic Stamp Company who acquired the Unicover business have failed elicit any reply.

5. As I set out on this quest I did not have a hard copy set of the stamps or stamps images that so interest me.

Where to look?

What if I was to set out to collect a copy of the set today from today's market?

My relationship, friendship with John, via the Internet, has been invaluable and enabled me to glean a very positive story of this stamp issue, an idea taken to fulfilment by his father.

John's father, James A Helzer is the entrepreneur who had started in the philatelic industry as a schoolboy, in Cheyenne, Wyoming supplying postage stamps, on envelopes, that were passed through the standard postal system to provide the collector with a "first day cover".

The stamp is certified as having been used on the first day of issue by the Post Offices' cancelling date stamp and the location of the posting. Over time the actual envelope was designed to enhance the image on the stamp and to provide an additional context to the stamp issue. From 1961 the business became known as Unicover Corporation and remained a family-run business. "Uni" implying a universal opportunity

Additional services were added such as the distribution of postage stamps within the United States of other nation's stamps. I must say I had not known that Great Britain, Australia, New Zealand and China, to name just four, availed of that service.

Historically, another company founded in 1929, Fleetwood Company of New York had become the largest supplier of first day covers in the world. In 1968 James A Helzer and Company, became an investor in Fleetwood and created the Fleetwood Division of Unicover. The business was endorsed by a collecting public. An example of the scope of the business was the sale of nearly 2million first day covers in 1969 of the first Moon landing.

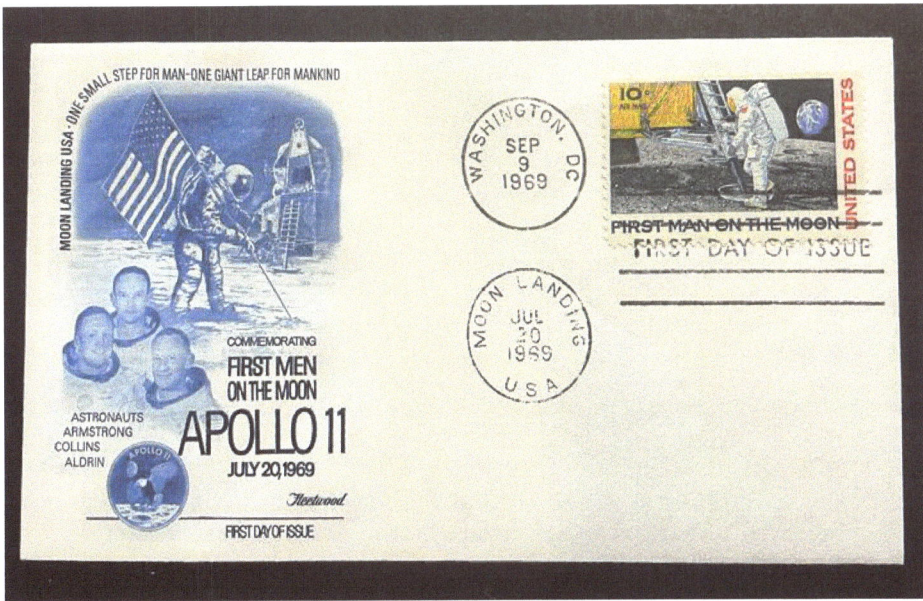

Seeking this cover on the Internet they are now obtainable for $100 or so.

Approaching the 50th anniversary of the end of World War Two and realising an opportunity for his company and the Republic of Marshall Islands (RMI), for whom Unicover provided Postal Administration Services, James Helzer put together a formidable team of specialists to record, via postage stamps, "**The one hundred epic events of World War II**".

Integrity was everything. The Military Historical Advisory Board appointed by James Helzer consisted of serving officers of rank, from the conflict, who understood "key events". The illustrators were proven war artists whose paintings were commissioned to be specific and unique to the Project.

The stamp images were to be slightly larger than normal Unicover issues and the designers were afforded a discretion as to portrait or landscape format, with the additional freedom to cover the main illustration over one, two or four actual stamps. The stamps would be printed in sheets of twelve or sixteen stamps of the same denomination. Every sheet to include the RMI Postal Services logo. The epic events to be recorded in sequence, 50 years after the event occurred.

At the conclusion of the commemoration of the 100 epic events Unicover took the opportunity record the project in a souvenir book.

Acquiring a set of the stamps and / or a set of the stamp images for this study.

I knew that I would find the set of stamp images within the Stampworld Internet site and the prices for the supply of all single images, both mint and used. I set out to complete for myself a virtual collection of the images that comprise the issue and use the opportunity to update via Wikipedia and other sources the current thinking of the history stories being illustrated.

What did surprise me was learning the fact that each of the Marshall Island / Unicover / Fleetwood images had been initially issued within a miniature sheet of multiple images. The standard configuration was 12 similar images within the miniature sheet. Better yet, some sheets included a montage of stamp images to tell the specific story; two stamps, perhaps or four. Depending on how the artist depicted the event the miniature sheet could consist of four sets of a montage of four images per event and sixteen stamps in the miniature sheet.

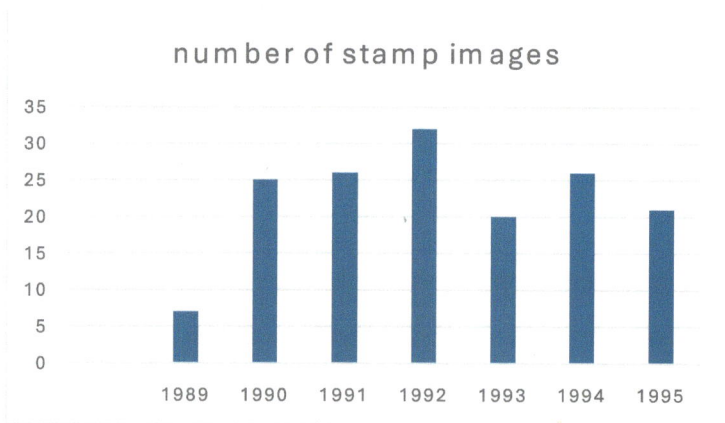

number of stamp images

The annual number of stamp images contained in the total set during the seven years of issue.

The project history of the set of stamps issued by the Unicover
Corporation, Fleetwood Division in 1995.

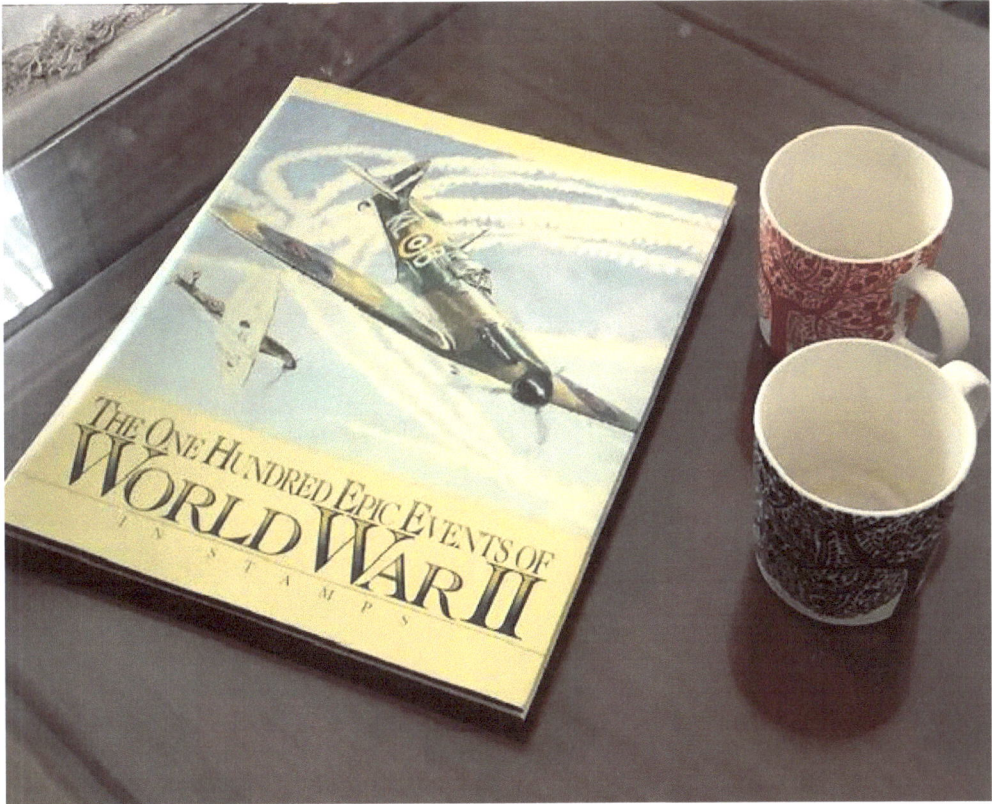

John Helzer has sent me one of the copies he has in his possession in
Longmont, Colorado. The front and back covers show an enlargement of the
action from Brian Sanders' illustration for the Battle of Britain event.

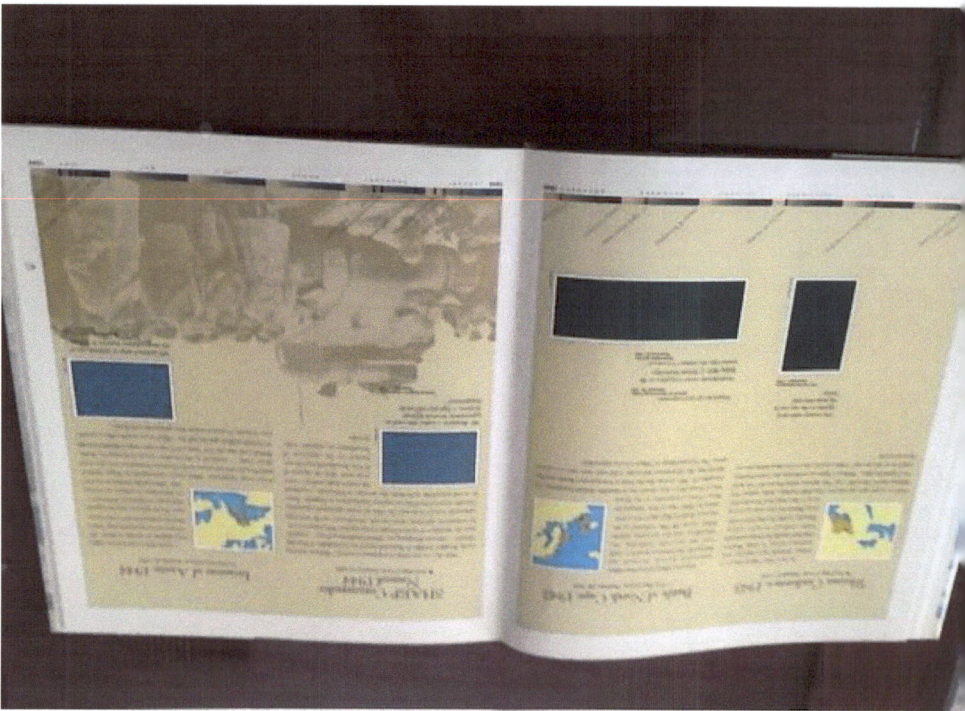

An internal 2-page view of the very glossy 1995, 64-page Unicover / Fleetwood 1995 publication "The One Hundred Epic Events of World War II".

Within the pages, as shown above, a map is included for every event and a space to mount the stamp example, (provided loose with the book). At the bottom of the illustrated page is a timeline of events during the year in question. In the example on this page, one of four semi-tone photographs from the invasion of Anzio 1944 we see Generals Eisenhower and Montgomery and other leaders inspecting the Anzio outcome.

The book was priced at US$150 in 1995 and included, in separate envelopes, a copy of every stamp and the appropriately cut mounts to handle all the formats within the set for the collector to mount them in the book.

I have been looking at Abebooks (second-hand Internet book dealer) and found that they have five copies of the (Fleetwood) 100 events of WW2 book. The price varies from $1 to $100 – but the narrative is very vague when it comes to stating (or not) whether the book contains any stamps. Another website (WorthPoint) has the book available with the stamps included – but it is not priced. (Noted January 2025)

STAMPS PACKAGE

THE ONE HUNDRED EPIC EVENTS OF
WORLD WAR II
IN STAMPS

World War II Stamps and Corresponding Mounts

Here is shown the "picking slip" for the stamps for inclusion in the Fleetwood book.

One Unicover Center · Cheyenne, Wyoming 82008-0001

Sourcing images of the set and determining prices, starting from scratch in the third quarter of 2024.

Anticipating a higher level of commerciality within the eBay price lists I looked there first. The images I obtained are mostly highlighted on coloured backing paper and totalled 5 miniature sheets, 15 singles, 5 doubles and 8 blocks of four images on later pages. The prices were (reasonable to my mind) approximately US$ 1 for each stamp, single, double, block of four or a sheet of 12 or 16 stamps.

I had recommended to me the Brookman Stamp Company, (of Vancouver, Washington), who specialise in Pacific Island stamp issues. Their prices are comparable to Stampworld and Bay and the on-line prices from the Deschamps catalogue. To test the credibility of Brookman I have ordered two interesting sheets from Brookman.

There is one vendor of items with the "100 epic events of WW2" set who is very active on eBay with the identity of "fishmull (4256)". *Fishmull* has sold via eBay since 1999. They offer the complete set, mint in blocks of four and including the selvedge commentary for $299 – which strikes me as reasonable.

Year of issue	Number of stamps	Cost of stamps US$ mint Brookman Stamps	Number of sheets 12 stamp sheets 16 stamp sheets	Cost of sheets US$ mint Brookman Stamps
1989	7	9.25	4	80.00
1990	25	28.50	10 plus 3	248.50
1991	26	30.75	7 plus 6	234.50
1992	32	38.75	21 plus 3	489.25
1993	20	28.75	9 plus 3	300.00
1994	26	45.75	16 plus 4	489.75
1995	21	62.50	13 plus 2	554.50
	157	$244.25	80 plus 21	$2,396.50

The annual number of sheets, containing 12 or 16 stamp images during the seven years of issue and the anticipated cost of acquiring from Brookman Stamp Company the full (mint) set.

But the sheet includes so much more for each issue.

The Philatelic items that describe the specific issue.

Colour print settings

Logo of RMI Postal Services

Miniature sheet / issue number

Marshall Islands 1991 – "The German invasion of Russia 1991.
Sheet W22. The 22nd image in the set.

Image number
Copyright confirmation   Selvedge commentary   Name of designer

I have not high-lighted the postal service fee. This changes a bit within the life of issue of the set. Logically, Unicover will have issued a low-level (pre-paid) service fee for postage within the Marshall Islands (29 / 30 cents) and 45 / 50 cents for overseas mail. There are one or two exceptions – The Battle of Iwo Jima and the Bomb upon Hiroshima have a $1 service fee. (US Dollars).

A 16-stamp sheet, the 26th sheet to be offered in the series, illustrating "Japanese attack Pearl Harbour".

Examples of the "first day of issue covers" postmarked Majuro, capital of the Marshall Islands Postal ServiceIslands postal service.

OFFICIAL FIRST DAY COVER

Battle of Taranto
November 11, 1940

FAIREY SWORDFISH 1

When an American airman first saw Great Britain's flimsy looking Fairey Swordfish 1 biplane, he exclaimed, "My God, you don't mean to say you fly those things!" But for all its fragile appearance, the Swordfish was an able craft and one of the most deadly of Great Britain's weapons. One of its most important qualities was its superior range. Carrying a single torpedo, the Swordfish could range over five hundred miles. British airmen nicknamed the plane "the Stringbag" after the shopping bag favored by many British housewives, since the plane was durable and versatile. The plane proved its capacity when twenty-one Swordfish brought the Italian Navy to its knees on the night of November 11, 1940 in Taranto Harbor.

Stamp Motif: Fairey Swordfish 1
Stamp Designer: Brian Sanders
First Day Cover Designer: Brian Sanders
Typographer: Jeff Bacon
Stamps Printed by: Unicover Corporation.
Printing Process: Offset

W17.FDC (4-2)

Marshall Islands 1990 : *Fairy Swordfish 1* [W17.fdc(4-2).   The text description on the back of the envelope reads :When an American airman first saw Great Britain's flimsy looking Fairey Swordfish1 biplane, he exclaimed, "my God, you don't mean to say you fly those things!" But for all its fragile appearance, the Swordfish was an able craft and one of the most deadly of Great Britain's weapons. One of its most important qualities was its superior range. Carrying a single torpedo, the Swordfish could range over five hundred miles. British Airmen nicknamed the plane "the Stringbag" after the shopping bag favoured by many British housewives, since the plane was durable and versatile. The plane proved its capacity when twenty-one Swordfish bought the Italian Navy to its knees on the night on November 11, 1940 in Taranto Harbour.

OFFICIAL FIRST DAY COVER

Japanese Attack Pearl Harbor
December 7, 1941

MARSHALL ISLANDS

JAPANESE ATTACK PEARL HARBOR

At 8 a.m. on the morning of December 7, 1941, Army Lt General Walter Short
heard from his quarters the distant rat-tat of gunfire and the muted roar of
aircraft engines. Within minutes a junior officer confirmed that Fort Shafter, 13
miles inland from Pearl Harbor, was under attack from enemy aircraft, as were
Wheeler and Hickam Fields and Kaneohe Naval Air Station. Chaos ruled the
day, for Short — fearful of sabotage — had ordered all aircraft to be kept in
wing-to-wing formations in the center of the unprotected airfields. Although
only a few aircraft survived the first attack, American pilots attempted to
pursue the enemy. But their valiant efforts could not stem the Japanese
onslaught, which had effectively destroyed 75 percent of America's Pacific-
based air power.

Stamp Motif: Army P-40, Marine SBD and F4F2
Stamp Designer: David Store
First Day Cover Designer: David Store
Typographer: Denise Fritze
Stamps Printed by: Unicover Corporation
Printing Process: Offset

W26.FDC (4-1)

Marshall Islands 1991 : *Japanese Attack Pearl Harbour* [W26.fdc(4-1]. At 8.a.m. on the morning of December 7, 1941, Army Lt General Walter Short heard from his quarters the distant rat-tat of gunfire and the muted roar of aircraft engines. Within minutes a junior officer confirmed that Fort Shorter, 13 miles inland from Pearl Harbour, was under attack from enemy aircraft, as were Wheeler and Hickam Fields and Kaneohe Naval Air Station. Chaos rules the day, for Short - fearful of sabotage – had ordered all aircraft to be kept in wing-to-wing formations in the centre of the unprotected airfields. Although only a few aircraft survived the first attack, American pilots attempted to pursue the enemy. But their valiant efforts could not stem the Japanese onslaught, which had effectively destroyed 75% of America's Pacific-based airpower.

So what makes this set "The best thematic set of postage stamps EVER" in my opinion?

My own bias towards military history on stamps.

The series illustrate world-history significant story told through visually compelling stamp images from beginning to end, complemented the historically exciting narratives (as seen at the 50th anniversary) of the event. I have added a commentary to each image to reflect the 2025 (80 year) perspective.

Consistently high-quality images with the acknowledgment to the individual artists who created them.

The integrity of the complete issue, over a 7-year period and being able to judge the significance of the events as seen by the Military Historical Panel who undertook the task of determining what were the epic events of WW2 at the 50-year anniversaries of those events and now after 80 years.

I particularly enjoy what I have called the selvedge commentary of participants in the event – an insight to how / what they felt at the time. It makes the event more alive.

Gaining an insight into the initiative, disciplined creativity and expertise of James A Helzer of the Unicover Corporation of Wyoming in creating the infrastructure to sustain his masterpiece.

The opportunity for me to update how history views the events 80 years after the event from up-to-date analysis.

There are three stamps issued with an anomaly, which were reprinted to correct a wrong spelling emphasising the quality controls exercised over the seven years of issue.

I have been able to present my study to the ACT Branch of the Military Historical Society of Australia to determine their enthusiasm for this study and the continuing interest in the history of the events via a "PowerPoint" media presentation.

My own acquisition of the stamp images for this study

My preferred format is the 12-stamp sheet. Sheets as well as the logical blocks of images (incorporating more than a single image to show one event, and singles are offered for sale on eBay by vendors who seem to specialise for the series.

In the pages below I show the 100 epic events as I accumulated over time.

W1 : Scott 239 : Marshall Islands 1989 : The Invasion of Poland, 1939█
Designer : David K Stone.
Selvedge commentary :
"When peace has been broken anywhere, the peace of all countries is in danger" —Franklin D Rooseveldt, September 3, 1939.

Germany invaded Poland on September 1, 1939, initiating World War II in Europe.
German forces broke through Polish defenses along the border and quickly advanced on Warsaw, the Polish capital. Hundreds of thousands of refugees, both Jewish and non-Jewish, fled the German advance hoping the Polish army

---

[1] The W designator shows the miniature sheet number within the set being discussed. The Scott Number refers to each stamps' identification within the Scott Catalogue, a major US postage stamp cataloguer.

could halt the German advance. But, after heavy shelling and bombing, Warsaw
surrendered to the Germans within a month of the German attack. Soviet forces
quickly annexed most of eastern Poland, while western Poland remained under
German occupation until 1945.

Britain and France, standing by their guarantee of Poland's border, declared war
on Germany on September 3, 1939. After the defeat of Polish forces, German
authorities began enforcing their racial policies in the occupied territories. They
required Jews to identify themselves by wearing white armbands with a blue Star
of David and conscripted them for forced-labor. They expelled hundreds of
thousands of Poles from their homes and settled more than 500,000 ethnic
Germans in their place.

Franklin Delano Roosevelt (January 30, 1882 – April 12, 1945), also known
as FDR, was the 32nd president of the United States, serving from 1933 until his
death in 1945. He is the longest-serving U.S. president, and the only one to have
served more than two terms. His initial two terms were centered on combating
the Great Depression, while his third and fourth saw him shift his focus to
America's involvement in World War II.

W2 : Scott 240 : Marshall Islands 1989 : Sinking of *HMS Royal Oak*, 1939.
Designer : Brian Sanders.
Selvedge commentary :
"On account of the obstacles imposed by the currents and the net barrages, this entry of a U-boat must be considered as a remarkable exploit of "professional skill and daring" - Winston S Churchill, October 17, 1939.

HMS Royal Oak (pennant number 91) was an aircraft carrier of the Royal Navy that was operated during the Second World War.
Royal Oak operated in some of the most active naval theatres of the Second World War. She was involved in the first aerial U-boat kills of the war, operations off Norway, the search for the German battleship *Bismarck*, and the Malta Convoys. *Royal Oak* survived several near misses and gained a reputation as a 'lucky ship'. She was torpedoed on 13 November 1941 by the German submarine *U-81* and sank the following day. One of her 1,488 crew members was killed. Her sinking was the subject of several inquiries, with investigators keen to know how the carrier was lost in spite of efforts to save the ship and tow her to the naval base at Gibraltar. They found that several design flaws contributed to the loss, which were rectified in new British carriers.

Winston Churchill was appointed First Lord of the Admiralty on 3 September 1939, the day that the United Kingdom declared war on Nazi Germany. He succeeded Neville Chamberlain as prime minister on 10 May 1940 and held the post until 26 July 1945. Out of office during the 1930s, Churchill had taken the lead in calling for British re-armament to counter the growing threat of militarism in Nazi Germany. Regarded as the most important of the Allied leaders during the first half of the War. Historians have long held Churchill in high regard as a victorious wartime leader who played an important role in defending Europe's liberal democracy against the spread of fascism. He has been consistently ranked both by scholars and the public as one of the top three greatest British prime ministers, often as the greatest prime minister in British history.

W3 : Scott 241 : Marshall Islands 1989 : The invasion of Finland, 1939 ■
Designer : David K Stone.
Selvedge commentary :
"We shall fight to the last old man and the last child. We shall burn our forests and houses,
destroy our cities and industries, and what we yield will be cursed by the scourge of God" –
Field Marshall Carl Gustav von Mannerhiem, Commander of the Finnish Army.

The Winter War was a war between the Soviet Union and Finland. It began with
a Soviet invasion of Finland on 30 November 1939, three months after the

---

[2] The vendor "Fishmull (4256)" distinguishes his items for sale via eBay mounting
their stamp images on a dark backing sheet. Several images are from this
source.

outbreak of World War II, and ended three and a half months later with
the Moscow Peace Treaty on 13 March 1940. Despite superior military strength,
especially in tanks and aircraft, the Soviet Union suffered severe losses and
initially made little headway. The League of Nations deemed the attack illegal and
expelled the Soviet Union from its organization.

The Soviets made several demands, including that Finland cede substantial
border territories in exchange for land elsewhere, claiming security reasons –
primarily the protection of Leningrad, 32 km from the Finnish border. When
Finland refused, the Soviets invaded. Most sources conclude that the Soviet
Union had intended to conquer all of Finland, and cite the establishment of
the puppet Finnish Communist government and the Molotov–Ribbentrop Pact's
secret protocols as evidence of this, while other sources argue against the idea of
a full Soviet conquest.

Carl Gustaf Emil Mannerheim (1867 – 1951) was a Finnish military commander,
aristocrat, and statesman. He served as the military leader of the Whites in
the Finnish Civil War (1918), as Regent of Finland (1918–1919), as commander-
in-chief of the Finnish Defence Forces during the period of World War
II (1939–1945), and as the sixth president of Finland (1944–1946). He became
Finland's only field marshal in 1933 and was appointed honorary Marshal of
Finland in 1942.

W4 : Scott 242 : The Battle of the River Plate 1939.

Designer : Brian Sanders.

Selvedge commentary :

"She (Graf Spee) was a sight to stir a seaman's heart — the lean strength of her fine flowing lines and her unbroken main deck sweeping abaft the after turret" – The New York Times, 1934.

"Attack at once by day or night"- British Commodore Harry Harwood, December 12, 1939.

The recognised craft are (1) *HMS Exeter*, (2) *HMS Alex*, (3) *Graf Spee* and (4) *HMNZS Achilles*.

Commodore Harwood received acclaim for his action against the German pocket battleship the 'Admiral Graf Spee' in which three British cruisers – HMS 'Ajax', 'Achilles' and 'Exeter' – engaged at the Battle of the River Plate on 13 December 1939, inflicting serious damage on the enemy. Trapped in Montevideo harbour, and falsely believing that a larger British force lay in wait, the German commander, Captain Hans Langsdorff, scuttled his ship on 17 December, providing the British with some welcome good news.

Admiral Sir Henry Harwood, KCB, OBE (1888 – 1950) won fame in the Battle of the River Plate during the Second World War. From December 1940 to April 1942, Rear-Admiral Harwood served as a Lord Commissioner of the Admiralty and Assistant Chief of Naval Staff.[2] In April 1942, Harwood was promoted to vice-admiral and Commander-in-Chief, Mediterranean Fleet, and flew his flag at HMS *Nile*.

W5 : Scott 246 and 247 : Marshall Islands 1990 :
The invasions of Denmark and Norway.

Designer : Brian Sanders.

Selvedge commentary :

Denmark : "*All civilization seems to have come to an end. I cannot understand how such terrible things can happen*" – King Haaken VII, April 1940.

Norway : "*I wish to spare my country further misfortune and misery*"- King Christian X, April 1940.

The invasion of Denmark

The German attack on Denmark on 9 April 1940 was a prelude to the invasion of Norway.

Denmark's strategic importance for Germany was limited. The invasion's primary purpose was to use Denmark as a staging ground for operations against Norway, and to secure supply lines to the forces about to be deployed there. An extensive network of radar systems was built in Denmark to detect British bombers bound for Germany.

The attack on Denmark was a breach of the non-aggression pact Denmark had signed with Germany less than a year earlier. The initial plan was to push Denmark to accept that German land, naval and air forces could use Danish bases, but Adolf Hitler subsequently demanded that both Norway and Denmark be invaded.

Denmark's military forces were inferior in numbers and equipment, and after a short battle were forced to surrender. After less than two hours of struggle, Danish Prime Minister Thorvald Stauning ended the opposition to the German attack, for fear that the Germans would bomb Copenhagen (København), as they had done with Warsaw during the invasion of Poland in September 1939. Due to communication difficulties, some Danish forces continued to fight, but after a further two hours, all opposition had stopped.

Lasting approximately six hours, the German ground campaign against Denmark was one of the shortest military operations of the Second World War.

The invasion of Norway

Poorly armed, neutral Norway became the first victim of Germany's Blitzkrieg in western Europe in April 1940.

Both the Allies and Germany ignored Norwegian neutrality. During the winter, Germany imported Swedish iron ore through the Norwegian port of Narvik. In response, Britain planned to lay mines along the Norwegian coast. British sailors also boarded the German ship Altmark in Norwegian waters.

Germany launched a full scale invasion on 9 April 1940. In a series of surprise attacks, 10,000 German troops seized the capital, Oslo, and the main ports. Although Allied efforts to intervene on land ended in failure, the invasion was costly for the German Navy. The new cruiser *Blücher* was sunk by Norwegian coastal guns at Oslo, and the scattered German ships were vulnerable to the Royal Navy, which scored a notable victory at Narvik. Further losses and damage to Germany's few modern warships were inflicted by Allied submarines and aircraft.

On land, the poorly equipped Allied troops were outnumbered and outgunned. By 2 May, most had been evacuated, although fighting continued at Narvik until the Germans had invaded France and Belgium, when it became urgent to save the remaining 24,000 Allied soldiers for use elsewhere. After destroying the port and railway, they were withdrawn. The tragic loss of the aircraft carrier *Glorious* ended the campaign.

The German Navy never really recovered from the losses sustained in Norway, which in the immediate aftermath prevented it from interfering with the evacuation of Dunkirk or supporting an invasion of Britain.

Haakon VII (1872 – 1957) was King of Norway from 18 November 1905 until his death in 1957. After the 1905 dissolution of the union between Sweden and Norway, he was offered the Norwegian crown. Following a November plebiscite, he accepted the offer and was formally elected king of Norway. As king, Haakon gained much sympathy from the Norwegian people. Although the Constitution of Norway vests the King with considerable executive powers, in practice Haakon confined himself to a representative and ceremonial role while rarely interfering in politics. When Norway was invaded by Nazi Germany in April 1940. Haakon rejected German demands to legitimise the Quisling regime's puppet government, vowing to abdicate rather than do so.

Christian X (1870 – 1947) was King of Denmark from 1912 until his death in 1947. During the German occupation of Denmark, Christian became a popular symbol of resistance, one of the most popular Danish monarchs of modern times.

W6 : Scott 248 : Marshall Islands 1990 : The Katyn Forest Massacre, 1940.
Designer : David K Stone.
Selvedge commentary :
"They had their hands tied up ... these Polish prisoners had perhaps tried to resist at the last moment" – General von Gersdorff, April 1952.

The Katyn massacre was a series of mass executions of nearly 22,000 defenceless Polish military and police officers, border guards, and intelligentsia prisoners of war carried out by the Soviet Union, specifically the NKVD (the Soviet secret police), at Stalin's order in April and May 1940. Though the killings also occurred in the Kalinin and Kharkiv NKVD prisons and elsewhere, the massacre is named after the Katyn forest, where some of the mass graves were first discovered by German Nazi forces. The massacre is qualified as a crime against humanity, crime against peace, war crime and Communist crime and according to a resolution of the Polish parliament, it bears the hallmarks of a genocide.

The order to execute captive members of the Polish officer corps was secretly issued by the Soviet Politburo led by Joseph Stalin. Of the total killed, about 8,000 were officers imprisoned during the 1939 Soviet invasion of Poland. The Polish Army officer class was representative of the multi-ethnic Polish state; the murdered included ethnic Poles, Ukrainians, Belarusians, and 700–900 Polish Jews. The government of Nazi Germany announced the discovery of mass graves in the Katyn Forest in April 1943.

Rudolf Christoph Freiherr von Gersdorff (1905 – 1980) was an officer in the German Army. As a Wehrmacht intelligence officer, he attempted to assassinate Adolf Hitler by suicide bombing on 21 March 1943; the plan failed when Hitler left early, but Gersdorff was undetected. That same month, soldiers from his unit discovered the mass graves of the Soviet-perpetrated Katyn massacre.

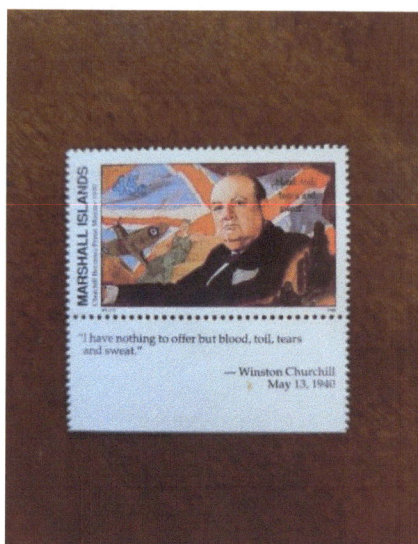

W7 : Scott 251 : Marshall Islands 1990 : Churchill becomes (UK) Prime Minister, 1940.

Designer : Shannon Stirnweis.

Selvedge commentary :

"I have nothing to offer but blood, toil, tears and sweat" – Winston Churchill May 13, 1940.

Sir Winston Spencer Churchill (1874 – 1965) was a British statesman, soldier, and writer who was twice Prime Minister of the United Kingdom, from 1940 to 1945 during the Second World War, and 1951 to 1955. Apart from 1922 to 1924, he was a Member of Parliament (MP) from 1900 to 1964 and represented a total of five constituencies. Ideologically an adherent to economic liberalism and imperialism, he was for most of his career a member of the Conservative Party, which he led from 1940 to 1955. He was a member of the Liberal Party from 1904 to 1924.

At the outbreak of the Second World War, he was re-appointed First Lord of the Admiralty. In May 1940, he became prime minister, succeeding Neville Chamberlain. Churchill formed a national government and oversaw British involvement in the Allied war effort against the Axis powers, resulting in victory in 1945. After the Conservatives' defeat in the 1945 general election, he became Leader of the Opposition. Amid the developing Cold War with the Soviet Union, he publicly warned of an "iron curtain" of Soviet influence in Europe and promoted European unity.

W8 : Scott 249-251 : Marshall Islands 1990 : The Invasion of the Low Countries
        1940. The bombing of Rotterdam and the invasion of Belgium.

Designer : Brian Sanders.

Selvedge commentary :

" . . . *the Germans did a monstrous thing. They resorted to a merciless bombardment on a
colossal scale of the open town of Rotterdam . . . one of the worst crimes in history*"- E N van
Kleffens, Netherlands Minister of Foreign Affairs.

"*Despite the solemn pledges made to the world, the German Empire . . . has attacked
Belgium, which has always remained loyal and neutral.*" Leopold III, King of Belgium.

In 1940, Rotterdam was subjected to heavy aerial bombardment by
the Luftwaffe during the German invasion of the Netherlands during the Second
World War. The objective was to support the German troops fighting in the city,
break Dutch resistance and force the Dutch army to surrender. Bombing began
at the outset of hostilities on 10 May and culminated with the destruction of the
entire historic city centre on 14 May, an event sometimes referred to as
the Rotterdam Blitz.

The psychological and the physical success of the raid, from the German
perspective, led the Oberkommando der Luftwaffe (OKL) to threaten to destroy
the city of Utrecht if the Dutch command did not surrender. The Dutch
surrendered in the late afternoon of 14 May and signed the capitulation early the
next morning.

The invasion of Belgium or Belgian campaign (10–28 May 1940), often referred to within Belgium as the 18 Days' Campaign formed part of the larger Battle of France, an offensive campaign by Germany during the Second World War. It took place over 18 days in May 1940 and ended with the German occupation of Belgium following the surrender of the Belgian Army.

On 10 May 1940, Germany invaded Luxembourg, the Netherlands, and Belgium under the operational plan *Fall Gelb* (Case Yellow). The Allied armies attempted to halt the German Army in Belgium, believing it to be the main German thrust. After the French had fully committed the best of the Allied armies to Belgium between 10 and 12 May, the Germans enacted the second phase of their operation, a break-through, or sickle cut, through the Ardennes, and advanced toward the English Channel. The German Army (*Heer*) reached the Channel after five days, encircling the Allied armies. The Germans gradually reduced the pocket of Allied forces, forcing them back to the sea. The Belgian Army surrendered on 28 May 1940, ending the battle.

Van Kleffens was appointed the Minister of Foreign Affairs in 1939, weeks before World War II began, and was part of the Dutch government in exile over that period. During the war he penned an account of the German invasion named *Juggernaut over Holland* which was circulated within the occupied territory, and he was also one of the original signatories of the Benelux union.

Van Kleffens held the position of foreign minister until the Schermerhorn–Drees cabinet of 1946. Following his resignation from the ministerial position (but not from the cabinet) van Kleffens became the Netherlands' representative on the United Nations Security Council, and in 1947 was appointed the ambassador to the United States. In 1950 he became the ambassador to Portugal, and was bestowed the title of Minister of State, a prestigious honour.

Leopold III (1901 – 1983) was King of the Belgians from 23 February 1934 until his abdication on 16 July 1951. At the outbreak of World War II, Leopold tried to maintain Belgian neutrality, but after the German invasion in May 1940, he surrendered his country, earning him much hostility, both at home and abroad.

W9 : Scott 252-253 : Marshall Islands 1940 : Evacuation of Dunkirk, 1940.
Designer : Brian Sanders.
Selvedge commentary :
"*We shall give you all that the Navy and the AirForce can do:* - Winston S Churchill to
General Lord Gort. May 27, 1940.
"*We must be careful not to assign to this deliverance the attributes of victory. Wars are not won*
by evacuations. 'But there was a victory inside this deliverance . . ." – Winston S Churchill
to the House of Commons, June 4, 1940.

The Dunkirk evacuation was the evacuation of more than 338,000 Allied soldiers
during the Second World War from the beaches and harbour of Dunkirk, in the
north of France, between 26 May and 4 June 1940. The operation commenced
after large numbers of Belgian, British, and French troops were cut off and
surrounded by German troops during the six-week Battle of France.

After Germany invaded Poland in September 1939, France and the British
Empire declared war on Germany and imposed an economic blockade.
The British Expeditionary Force (BEF) was sent to help defend France. After
the Phoney War of October 1939 to April 1940, Germany invaded Belgium,
the Netherlands, and France on 10 May 1940. Three panzer corps attacked
through the Ardennes and drove northwest to the English Channel. By 21 May,
German forces had trapped the BEF, the remains of the Belgian forces, and
three French field armies along the northern coast of France. BEF
commander General Viscount Gort immediately saw evacuation across the
Channel as the best course of action, and began planning a withdrawal to
Dunkirk, the closest good port.
Late on 23 May, the halt order was issued by *Generaloberst* Gerd von Rundstedt,
commander of Army Group A. Adolf Hitler approved this order the next day,
and had the German High Command send confirmation to the front. Attacking
the trapped BEF, French, and Belgian armies was left to the *Luftwaffe* until the
order was rescinded on 26 May. This gave Allied forces time to construct

defensive works and pull back large numbers of troops to fight the Battle of Dunkirk. From 28 to 31 May, in the siege of Lille, the remaining 40,000 men of the French First Army fought a delaying action against seven German divisions, including three armoured divisions.

By the end of the eighth day, 338,226 persons had been rescued by a hastily assembled fleet of over 800 vessels. Many troops were able to embark from the harbour's protective mole onto 39 British Royal Navy destroyers, four Royal Canadian Navy destroyers, at least three French Navy destroyers, and a variety of civilian merchant ships. Others had to wade out from the beaches, waiting for hours in shoulder-deep water. Some were ferried to the larger ships by what became known as the Little Ships of Dunkirk, a flotilla of hundreds of merchant marine boats, fishing boats, pleasure craft, yachts, and lifeboats called into service from Britain. The BEF lost 68,000 soldiers during the French campaign and had to abandon nearly all of its tanks, vehicles, and equipment. In his "We shall fight on the beaches" speech on 4 June to the House of Commons, British Prime Minister Winston Churchill called the event "a colossal military disaster", saying "the whole root and core and brain of the British Army" had been stranded at Dunkirk and seemed about to perish or be captured. He hailed their rescue as a "miracle of deliverance". Churchill also reminded the country that "we must be very careful not to assign to this deliverance the attributes of a victory. Wars are not won by evacuations."

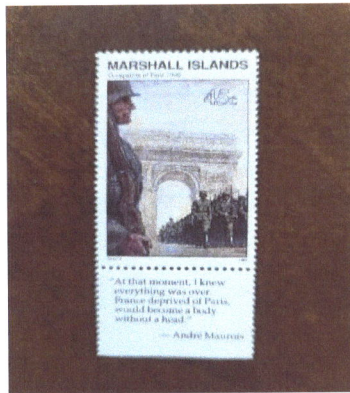

W10 : Scott 254 : Marshall Islands 1990 : The occupation of Paris, 1940.
Designer : David K Stone.
Selvedge commentary :
"At that moment I knew everything was over. France deprived of Paris would become a body without a head". André Maurois.

Paris started mobilizing for war in September 1939, when Nazi Germany and the Soviet Union attacked Poland, but the war seemed far away until 10 May 1940, when the Germans attacked France and quickly defeated the French army. The French government departed Paris on 10 June, and the Germans occupied the city on 14 June. During the occupation, the French government moved to Vichy, and Paris was governed by the German military and by French officials approved by the Germans. For Parisians, the occupation was a series of frustrations, shortages and humiliations. A curfew was in effect from nine in the evening until five in the morning; at night, the city went dark. Rationing of food, tobacco, coal and clothing was imposed from September 1940. Every year the supplies grew the more scarce and the prices higher. A million Parisians left the city for the provinces, where there was more food and fewer Germans. The French press and radio contained only German propaganda.

Jews in Paris were forced to wear the yellow Star of David badge, and were barred from certain professions and public places. On 16–17 July 1942, 13,152 Jews, including 4,115 children, were rounded up by the French police, on orders of the Germans, and were sent to the Auschwitz concentration camp. The first demonstration against the occupation, by Paris students, took place on 11 November 1940. As the war continued, anti-German clandestine groups and networks were created, some loyal to the French Communist Party, others to

General Charles de Gaulle in London. They wrote slogans on walls, organized an underground press, and sometimes attacked German officers. Reprisals by the Germans were swift and harsh.

André Maurois: a French biographer, novelist, and essayist, best known for biographies that maintain the narrative interest of novels.

W11 : Scott 255 : Marshall Islands 1990 : Mers-el-Kabir, 1940.
Designer : Shannon Stirnweis, a well-known American illustrator in the 1970s.
Selvedge commentary :
"I have the orders ... to use whatever force may be necessary to prevent your ships from falling into German or Italian hands" – Vice-Admiral Sir James Somerville to French Admiral Marcel-Bruno Gensoul. July 3, 1940.

The attack on Mers-el-Kébir on 3 July 1940, during the Second World War, was a British naval attack on neutral French Navy ships at the naval base at Mers El Kébir, near Oran, on the coast of French Algeria. The attack was the main part of Operation Catapult, a British plan to neutralise or destroy neutral French ships to prevent them from falling into German hands after the Allied defeat in the Battle of France. The British bombardment of the base killed 1,297 French servicemen, sank a battleship and damaged five other ships, for a British loss of five aircraft shot down and two crewmen killed. The attack by air and sea was

conducted by the Royal Navy, after France had signed armistices
with Germany and Italy, coming into effect on 25 June.

After he had destroyed the French Battle fleet, Admiral Somerville played an
important role in the pursuit and sinking of the German battleship *Bismarck*.
Marcel-Bruno Gensoul never commented upon the event.

W12 : Scott 256 : Marshall Islands 1990 : The Burma Road conflicts, 1940-1945.
Designer : David K Stone.
Selvedge commentary :
"We got run out of Burma and it is humiliating . . . we ought to find out what caused it, go back and retake it". Lieutenant General Joseph Stilwell.

The Burma campaign was a series of battles fought in the British colony of Burma. It was part of the South-East Asian theatre of World War II and primarily involved forces of the Allies (mainly from the British Empire and the Republic of China, with support from the United States) against the invading forces of the Empire of Japan. Imperial Japan was supported by the Thai Phayap Army, as well as two collaborationist independence movements and armies. In 1942 and 1943, the international Allied force in British India launched several failed offensives to retake lost territories. Fighting intensified in 1944, and British Empire forces peaked at around 1 million land and air forces. These forces were drawn primarily from British India, with British Army forces (equivalent to eight regular infantry divisions and six tank regiments), 100,000 East and West African colonial troops, and land and air forces from other Dominions and Colonies. These additional forces allowed the Allied recapture of Burma in 1945. As the Imperial Japanese Army swept across China and South Asia at World War II's outset, closing all of China's seaports, more than 200,000 Chinese labourers embarked on a seemingly impossible task: to cut a 700-mile overland route -- the Burma Road -- from the southwest Chinese city of Kunming to Lashio, Burma. But when Burma fell in 1942, the Burma Road was severed. As the first step of the Allied offensive toward Japan, American general Joseph Stilwell reopened it, while, at the same time, keeping China supplied by air-lift from India and simultaneously driving the Japanese out of Burma.

Joseph Stilwell (1883-1946) was a United States Army general during World War II. Stilwell was made the Chief of Staff to the Chinese Nationalist Leader, Chiang

Kai-shek. He spent the majority of his tenure striving for a 90-division army trained by American troops, using American lend-lease equipment, and fighting to reclaim Burma from the Japanese. His efforts led to friction with Chiang, who viewed troops not under his immediate control as a threat, and who saw the Chinese communists as a greater rival than Japan. An early American popular hero of the war for leading a column walking out of Burma pursued by the victorious Imperial Japanese Armed Forces.

W13 : Scott 257-260 : Marshall Islands 1990 : Lend Lease, Destroyers to Great Britain.

Designer : David K Stone.

Selvedge commentary :

"The moral value of this fresh aid from your Government and people at this critical time will be very great and widely felt" – Winston S Churchill, August 15, 1940.

"It is an epochal and far-reaching act of preparation for continental defence in the face of grave danger" – Franklin D Roosevelt, September 3, 1940.

The destroyers-for-bases deal was an agreement between the United States and the United Kingdom on September 2, 1940, according to which 50 *Caldwell*, *Wickes*, and *Clemson*-class US Navy destroyers were transferred to the Royal Navy from the US Navy in exchange for land rights on British possessions.

Generally referred to as the "twelve hundred-ton type" (also known as "flush-deck", or "four-pipers" after their four funnels), the destroyers became the British Town class and were named after towns common to both countries. US President Franklin Roosevelt used an executive agreement, which does not require congressional approval. However, he came under heavy attack from antiwar Americans, who pointed out that the agreement violated the Neutrality Acts.

The images show :
HMS Georgetown ex-USS Maddox,
HMS Banff ex-USCCC Saranak,
HMS Buxton ex-USS Edwards, and
HMS Rockingham ex-USS Swasey.

HMS Georgetown participated in operation *"Bowery"*, escorting the American aircraft carrier USS Wasp in May 1942 on her second reinforcement of the spitfire strength on the island of Malta. In September 1942, she transferred to the Royal Canadian Navy tor convoy escort duties in the western Atlantic. returned, to the United Kingdom in December 1943, she Joined the Reserve Fleet and in August 1944 was turned over to the Soviet Navy.

HMS Banff, commissioned in 1941 was used extensively as a convey escort, prior to return to the US Coastguard service. She sailed from Portsmouth on 24th January 1946. The ship was Paid-Off at Boston on 27th February and resumed Coastguard service as USS Tampa.

HMS Buxton was commissioned in the Royal Navy 8 October 1940. The overage destroyer served in Canadian waters briefly as the U-boat war intensified. She was then allocated to 6th Escort Group, Western Approaches Command. In October 1943, when newer escorts were available, she was lent to the Royal Canadian Navy, and stationed at Digby, Nova Scotia, until the end of 1944. She was finally paid off on 2 June 1945.

HMS Rockingham had a November 1940 commissioning but was but lost in September 1944 when she struck a mine and sank in tow about 30 nautical miles south-east of Aberdeen in position, while acting as target ship for aircraft training.

W14 : Scott 261-264 : Marshall Islands 1990 : The Battle of Britain 1940.
Designer : Brian Sanders.
Selvedge commentary :
"The Battle of Britain is about to begin. Members of the Air Force: The fate of generations lies in your hands" – RAF Air Marshal Hugh Dowding, August 8, 1940.
"Never in the field of human conflict has so much been owed by so many to so few" – Winston S Churchill, August 20, 1940.

The recognised weapons are (1) *Supermarine Spitfire Mk.1*, (2) *Hawker Hurricane Mk.1,* (3) *Messerschmidt Bf 109E* and (4) *Junkers Ju 87B-2.*

The Battle of Britain was a military campaign of the Second World War, in which the Royal Air Force (RAF) and the Fleet Air Arm (FAA) of the Royal Navy defended the United Kingdom against large-scale attacks by Nazi Germany's air force, the Luftwaffe. It was the first major military campaign fought entirely by air forces. The British officially recognise the battle's duration as being from 10 July until 31 October 1940, which overlaps the period of large-scale night attacks known as the Blitz, that lasted from 7 September 1940 to 11 May 1941.

Air Chief Marshal Hugh Dowding, 1st Baron Dowding, (1882 –1970) was a senior officer in the Royal Air Force. He was Air Officer Commanding RAF Fighter Command during the Battle of Britain and is generally credited with playing a crucial role in Britain's defence, and hence, the defeat of Operation Sea Lion, Adolf Hitler's plan to invade Britain.

W15 : Scott 265 : Marshall Islands 1990 : The Tri-Partite Pact, 1940.
Designer : David K Stone.
Selvedge commentary :
"The announcement makes clear to all a relationship that has long existed". Cordell Hull,
US secretary of State, September 27, 1940

The Tripartite Pact was the agreement between Germany, Italy, and Japan signed
in Berlin on 27 September 1940 by, respectively, Joachim von
Ribbentrop, Galeazzo Ciano, and Saburō Kurusu (in that order) and in the
presence of Adolf Hitler. It was a defensive military alliance that was eventually
joined by Hungary (20 November 1940), Romania (23 November
1940), Slovakia (24 November 1940), Bulgaria (1 March 1941),
and Yugoslavia (25 March 1941). Yugoslavia's accession provoked a *coup d'état* in
Belgrade two days later. Germany, Italy, and Hungary responded by invading
Yugoslavia. The resulting Italo-German client state, known as the Independent
State of Croatia, joined the pact on 15 June 1941.
The Tripartite Pact was, together with the Anti-Comintern Pact and the Pact of
Steel, one of a number of agreements between Germany, Japan, Italy, and other
countries of the Axis Powers governing their relationship.
The Tripartite Pact formally allied the Axis Powers with one another, and it was
directed primarily at the United States. Because of the long distance between
Japan and the two European Powers, the pact recognized two different regions
that were to be under Axis rule. Japan recognized "the leadership of Germany
and Italy in the establishment of a new order in Europe". In return, Germany
and Italy recognized Japan's right to establish a new order "in Greater East
Asia". But the pact's practical effects were limited since the Italo-German and
Japanese operational theatres were on opposite sides of the world, and the high
contracting powers had disparate strategic interests. As such, the Axis was only

ever a loose alliance. Its defensive clauses were never invoked, and signing the agreement did not oblige its signatories to fight a common war *per se.*

Cordell Hull (October 2, 1871 – July 23, 1955) was an American politician from Tennessee and the longest-serving U.S. Secretary of State, holding the position for 11 years (1933–1944) in the administration of President Roosevelt during most of World War II. Before that appointment, Hull represented Tennessee for two years in the United States Senate and 22 years in the House of Representatives.

W16 : Scott 266 : Marshall Islands 1990 : Re-election of President Franklin Roosevelt, 1940.

Designer : David K Stone.

Selvedge commentary : *"We cannot accept that war must be forever a part of man's destiny"* – Franklin D Roosevelt, November 2, 1940.

Franklin Delano Roosevelt (1882 – 1945), commonly known by his initials FDR, was an American politician who served as the 32nd president of the United States from 1933 until his death in 1945. The longest serving U.S. president, he is the only president to have served more than two terms. His initial two terms were centered on combating the Great Depression, while his third and fourth saw him shift his focus to America's involvement in World War II.

Following the Japanese attack on Pearl Harbor on December 7, 1941, Roosevelt obtained a declaration of war on Japan. After Germany and Italy declared war on the U.S. on December 11, 1941, the U.S. Congress approved additional declarations of war in return. He worked closely with other national leaders in leading the Allies against the Axis powers. Roosevelt supervised the mobilization of the American economy to support the war effort and implemented a Europe first strategy. He also initiated the development of the first atomic bomb and worked with the other Allied leaders to lay the groundwork for the United Nations and other post-war institutions, even coining the term "United Nations". Roosevelt won re-election in 1944 but died in 1945 after his physical health seriously and steadily declined during the war years. Since then, several of his actions have come under criticism, such as his ordering of the internment of Japanese Americans. Nonetheless, historical rankings consistently place him among the three greatest American presidents.

W17 : Scott 267-270 : Marshall Islands 1990 : Battle of Taranto,1940.
Designer : Brian Sanders.
Selvedge commentary : *"It proved : once and for all time that in the Fleet Air Arm the Navy has its most devestating weapon"* - Admiral Andrew Cunningham, British Naval Commander.
"By this single stroke the balance of naval powerin the Mediterranean was decively altered" - Winston S Churchill.

The Battle of Taranto took place on the night of 11/12 November 1940 during the Second World War between British naval forces, under Admiral Andrew Cunningham, and Italian naval forces, under Admiral Inigo Campioni. The Royal Navy launched the first all-aircraft ship-to-ship naval attack in history, employing 21 Fairey Swordfish biplane torpedo bombers from the aircraft carrier HMS *Illustrious* in the Mediterranean Sea.

The attack struck the battle fleet of the *Regia Marina* at anchor in the harbour of Taranto, using aerial torpedoes despite the shallowness of the water. The success of this attack augured the ascendancy of naval aviation over the big guns of battleships. According to Admiral Cunningham, "Taranto, and the night of 11–12 November 1940, should be remembered forever as having shown once and for all that in the Fleet Air Arm the Navy has its most devastating weapon."

Admiral of the Fleet Andrew Browne Cunningham, 1st Viscount Cunningham of Hyndhope, (1883 – 1963) was widely known by his initials, "ABC". In the Second World War, as Commander-in-Chief, Mediterranean Fleet, Cunningham led British naval forces to victory in several critical Mediterranean naval battles. These included the attack on Taranto in 1940.

W18 : Scott 271-274 : Marshall Islands 1991 : President Roosevelt's Four
Freedoms speech 1941.

Designer : Howard Koslow.

Selvedge commentary :

". . . we look forward to a world founded upon four essential human freedoms . . ."
". . . freedom of speech and expression . . . freedom to worship God in his own way . . . freedon
from want . . . freedom from fear". FDR During the speech articulated on Monday,
January 6, 1941.

The State of the Union speech before Congress was largely about the national
security of the United States and the threat to other democracies from world
war. In the speech, he made a break with the long-held tradition of United States
non-interventionism. He outlined the U.S. role in helping allies already engaged
in warfare, especially Great Britain and China.

The ideas enunciated in the Roosevelt's Four Freedoms were the foundational
principles that evolved into the Atlantic Charter declared by Winston Churchill
and FDR in August 1941; the United Nations Declaration of January 1, 1942;
President Roosevelt's vision for an international organization that became the
United Nations after his death; and the Universal Declaration of Human Rights
adopted by the United Nations in 1948 through the work of Eleanor Roosevelt
(https://www.fdrlibrary.org/four-freedoms).

W19 : Scott 275 : Marshall Islands 1991 : The Battle of Beda Fomm, 1941.
Designer : Brian Sanders.
Selvedge commentary :
"The events in Libya are only part . . . of the story of the decline and fall of the Italian Empire"- Winston S Churchill, February 9, 1941.

The Battle of Beda Fomm took place following the rapid British advance during Operation Compass (9 December 1940 – 9 February 1941). The Italian 10th Army was forced to evacuate Cyrenaica, the eastern province of Libya. In late January, the British learned that the Italians were retreating along the Litoranea Balbo from Benghazi. The 7th Armoured Division (Major-General Sir Michael O'Moore Creagh) was dispatched to intercept the remnants of the 10th Army by moving through the desert, south of the Jebel Akhdar (Green Mountain) via Msus and Antelat as the 6th Australian Division pursued the Italians along the coast road, north of the jebel. The terrain was hard going for the British tanks and Combeforce (Lieutenant-Colonel John Combe), a flying column of wheeled vehicles, was sent ahead across the chord of the jebel.

Late on 5 February, *Combeforce* arrived at the Via Balbia south of Benghazi and set up roadblocks near Sidi Saleh, about 30 mi (48 km) south-west of Antelat and 20 mi (32 km) north of Ajedabia. The leading elements of the 10th Army arrived thirty minutes after the British, who sprung the ambush. Next day the Italians attacked to break through and continued their attacks into 7 February. With British reinforcements arriving and the Australians pressing down the road from Benghazi, the 10th Army surrendered later that day. Between Benghazi to Agedabia, the British took 25,000 prisoners, captured 107 tanks and 93 guns out of the totals for Operation Compass of 133,298 men, 420 tanks and 845 guns.

On 9 February, Churchill ordered the advance to stop and troops to be dispatched to Greece to take part in the Greco-Italian War; a German attack through Macedonia was thought imminent. The British in Libya were unable to continue beyond El Agheila anyway, because of vehicle breakdowns, exhaustion and the effect of the much longer distance from the supplies in Egypt. A few thousand men of the 10th Army escaped the disaster in Cyrenaica but the 5th Army in Tripolitania had four divisions. The Sirte, Tmed Hassan and Buerat strongholds were reinforced from Italy, which brought the 10th and 5th armies up to about 150,000 men. German reinforcements were sent to Libya to form a blocking detachment under Directive 22 (11 January), these being the first units of the Afrika Korps (Generalleutnant Erwin Rommel).

W20 : Scott 276-277 : Marshall Islands 1991 :
Germany invades the Balkans, 1941.
(1) Invasion of Greece, (2) Invasion of Yugoslavia.

Designer : Brian Sanders.

Selvedge commentary : *"We are a small people, but it is the fate of certain nations that Providence gives them the mission to fight against power of evil and darkness"* – Greek Premier Korizis.

"The barbaric invasion of Yugoslavia . . . is but another chapter in the present planned movement of attempted world conquest and domination" – Cordell Hull, US Secretary of State

The recognised weapons of war are the German *Focke-Wulf Fw 190* aircraft and the *Panzer III tank.*

The German invasion of Greece, were the attacks
on Greece by Italy and Germany during World War II. The Italian invasion in October 1940, which is usually known as the Greco-Italian War, was followed by the German invasion in April 1941. German landings on the island of Crete (May 1941) came after Allied forces had been defeated in mainland Greece. These battles were part of the greater Balkans Campaign of the Axis powers and their associates.

Following the Italian invasion on 28 October 1940, Greece, with British air and material support, repelled the initial Italian attack and a counter-attack in March 1941. When the German invasion, known as Operation Marita, began on 6 April, the bulk of the Greek Army was on the Greek border with Albania, then a vassal of Italy, from which the Italian troops had attacked. German troops invaded from Bulgaria, creating a second front. Greece received a small reinforcement from British, Australian and New Zealand forces in anticipation of the German attack.

The invasion of Yugoslavia, also known as the April War or Operation 25, was a German-led attack on the Kingdom of Yugoslavia by the Axis powers which began on 6 April 1941 during World War II. The order for the invasion was put forward in "Führer Directive No. 25", which Adolf Hitler issued on 27 March 1941, following a Yugoslav coup d'état that overthrew the pro-Axis government.

The invasion commenced with an overwhelming air attack on Belgrade and facilities of the Royal Yugoslav Air Force (VVKJ) by the Luftwaffe (German Air Force) and attacks by German land forces from southwestern Bulgaria. These attacks were followed by German thrusts from Romania, Hungary and the Ostmark (modern-day Austria, then part of Germany). Italian forces were limited to air and artillery attacks until 11 April, when the Italian army attacked towards Ljubljana (in modern-day Slovenia) and through Istria and Lika and down the Dalmatian coast. On the same day, Hungarian forces entered Yugoslav Bačka and Baranya, but like the Italians they faced practically light resistance. A Yugoslav attack into the northern parts of Italian-controlled Albania met with initial success, but was inconsequential due to the collapse of the rest of the Yugoslav forces and was ultimately repelled into Dalmatia.

Alexandros Koryzis (1885 – 1941) was a Greek politician who served briefly as the prime minister of Greece in 1941.

W21 : Scott 278-281Marshall Islands 1991 : The sinking of the Bismark, 1941.
Designer : Brian Sanders.
Selvedge commentary :
"Ship unmaneuverable. We shall fight to the last shell'- Admiral Günther Lütjens of the
Bismark".
*"The Bismark had put up a most gallant fight against impossible odds . . . she went down with
her colours still flying"* – Admiral John Tovey, Commander in Chief of the Home
Fleet.

The last battle of the German battleship *Bismarck* took place in the Atlantic
Ocean approximately 300 nautical miles (560 km; 350 mi) west of Brest, France,
on 26–27 May 1941 between the German battleship *Bismarck* and naval and air
elements of the British Royal Navy. Although it was a decisive action
between capital ships, it has no generally accepted name. It was the culmination
of Operation Rheinübung where the attempt of two German ships to disrupt the
Atlantic Convoys to the United Kingdom failed with the scuttling of
the *Bismarck*.

The last battle consisted of four main phases. The first phase late on the 26th
consisted of air strikes by torpedo bombers from the British aircraft carrier *Ark
Royal*, which disabled *Bismarck*'s steering gear, jammed her rudders in a turning
position and prevented her escape. The second phase was the shadowing and
harassment of *Bismarck* during the night of 26/27 May by British and
Polish destroyers, with no serious damage to any ship. The third phase on the
morning of 27 May was an attack by the British battleships *King George
V* and *Rodney* supported by the heavy cruisers *Norfolk* and *Dorsetshire*. After about
100 minutes of fighting, *Bismarck* was sunk by the combined effects of shellfire,

torpedo hits and deliberate scuttling. On the British side, *Rodney* was lightly damaged by near-misses and by the blast effects of her own guns. British warships rescued 110 survivors from *Bismarck* before being obliged to withdraw because of an apparent U-boat sighting, leaving several hundred men to their fate. A U-boat and a German weathership rescued five more survivors. In the final phase, the withdrawing British ships were attacked the next day on 28 May by aircraft of the *Luftwaffe*, resulting in the loss of the destroyer HMS *Mashona*. Johann Günther Lütjens (1889 – 1941) was the German admiral whose military service spanned more than 30 years and two world wars. Lütjens is best known for his actions during World War II and his command of the battleship *Bismarck* during her foray into the Atlantic Ocean in 1941. He was killed in action during the last battle of the battleship Bismarck.

John Cronyn Tovey, 1st Baron Tovey (1885 – 1971), was sometimes known as Jack Tovey, During the First World War he commanded the destroyer HMS *Onslow* at the Battle of Jutland and then commanded the destroyer *Ursa* at the Second Battle of Heligoland Bight. During the Second World War he initially served as Second-in-Command of the Mediterranean Fleet in which role he commanded the Mediterranean Fleet's Light Forces (i.e. cruisers and destroyers). He then served as Commander-in-Chief of the Home Fleet and was responsible for orchestrating the pursuit and destruction of the *Bismarck*.

W22 : Scott 282 : Marshall Islands 1991 : The German Invasion of Russia, 1941.
Designer : David K Stone.
Selvedge commentary :
"You will regret that you attacked the Soviet Union. You will pay dearly for this yet."
- V G Dekenozov, Soviet Ambassador to Germany.

The recognised weapons are German Panzer III tanks.

Operation Barbarossa was the invasion of the Soviet Union by Nazi
Germany and many of its Axis allies, starting on Sunday, 22 June 1941 It was the
largest and costliest land offensive in human history, with around 10 million
combatants taking part, and over 8 million casualties by the end of the operation.

The attack put into action Nazi Germany's ideological goals of
eradicating communism, and conquering the western Soviet Union
to repopulate it with Germans. The German *Generalplan Ost* aimed to use some
of the conquered people as forced labour for the Axis war effort while acquiring
the oil reserves of the Caucasus as well as the agricultural resources of various
Soviet territories, including Ukraine and Byelorussia. Their ultimate goal was to
create more *Lebensraum* (living space) for Germany, and the eventual
extermination of the native Slavic peoples by mass deportation
to Siberia, Germanisation, enslavement, and genocide.

In the two years leading up to the invasion, Nazi Germany and the Soviet
Union signed political and economic pacts for strategic purposes. Following
the Soviet occupation of Bessarabia and Northern Bukovina, the German High
Command began planning an invasion of the Soviet Union in July 1940. Over
the course of the operation, over 3.8 million personnel of the Axis powers—the
largest invasion force in the history of warfare—invaded the western Soviet
Union, along a 2,900-kilometer (1,800 mi) front, with 600,000 motor vehicles

and over 600,000 horses for non-combat operations. The offensive marked a massive escalation of World War II, both geographically and with the Anglo-Soviet Agreement, which brought the USSR into the Allied coalition.

I have been unable to learn any more about V G Dekenozov through the Internet.

W23 : Scott 283-284 : Marshall Islands 1991 : The Atlantic Charter 1941.
Designer : Brian Sanders.
Selvedge commentary :
"...it is difficult...to agree to a world peace which would give to Nazism dominion over large numbers of conquered nations" – Franklin D Roosevelt.
"...it symbolizes...of the good forces of the world against the evil forces..." Winston S Churchill.

The war ships illiustrated are the USS Augusta and HMS Prince of Wales.

The Atlantic Charter was a statement issued on 14 August 1941 that set out the American and British goals for the world after the end of World War II, months before the US officially entered the war. The joint statement, later dubbed the Atlantic Charter, outlined the aims of the United States and the United Kingdom for the postwar world as follows: no territorial aggrandizement, no territorial changes made against the wishes of the people (self-determination), restoration of self-government to those deprived of it, reduction of trade restrictions, global co-operation to secure better economic and social conditions for all, freedom from fear and want, freedom of the seas, abandonment of the use of force, and disarmament of aggressor nations. The charter's adherents signed the Declaration by United Nations on 1 January 1942, which was the basis for the modern United Nations.
The charter inspired several other international agreements and events after the war. The dismantling of the British Empire, the formation of NATO, and the General Agreement on Tariffs and Trade all derived from the Atlantic Charter.

USS *Augusta* (CL/CA-31) was a *Northampton*-class cruiser of the United States Navy, notable for service as a headquarters ship during Operation Torch, Operation Overlord, and Operation Dragoon, and for her occasional use as a presidential flagship.

HMS *Prince of Wales* was a *King George V*-class battleship of the Royal Navy. Despite being sunk less than a year after she was commissioned, she was involved in several key actions of the Second World War, including the May 1941 Battle of the Denmark Strait where she scored three hits on the German battleship *Bismarck*, forcing *Bismarck* to abandon her raiding mission and head to port for repairs.

W24: Scott 285 : Marshall Islands 1991 : The siege of Moscow, 1941.
Designer : David K Stone.
Selvedge commentary :
"The only way Hitler will ever see the kremlin is in a photograph"- Solomon A
Lozovsky, Assistant People's Commissar for Foreign Affairs.

The Battle of Moscow was the military campaign that consisted of two periods
of strategically significant fighting on a 600 km sector of the Eastern
Front during World War II, between September 1941 and January 1942. The
Soviet defensive effort frustrated Hitler's attack on Moscow, the capital and
largest city of the Soviet Union. Moscow was one of the primary military and
political objectives for Axis forces in their invasion of the Soviet Union.

The German Strategic Offensive, named Operation Typhoon, called for
two pincer offensives, one to the north of Moscow against the Kalinin Front by
the 3rd and 4th Panzer Armies, simultaneously severing the Moscow–Leningrad
railway, and another to the south of Moscow Oblast against the Western
Front south of Tula, by the 2nd Panzer Army, while the 4th Army advanced
directly towards Moscow from the west.

Initially, the Soviet forces conducted a strategic defence of the
Moscow Oblast by constructing three defensive belts, deploying newly
raised reserve armies, and bringing troops from the Siberian and Far Eastern
Military Districts. As the German offensives were halted, a Soviet
strategic counter-offensive and smaller-scale offensive operations forced the
German armies back to the positions around the cities
of Oryol, Vyazma and Vitebsk, and nearly surrounded three German armies. It
was a major setback for the Germans, and the end of their belief in a swift
German victory over the USSR. As a result of the failed offensive, Field
Marshal Walther von Brauchitsch was dismissed as supreme commander of
the German Army, with Hitler replacing him in the position.

Solomon Abramovich Lozovsky (1878–1952) was a prominent Communist and Bolshevik revolutionary, a high-ranking official in the Soviet government, including as a Presidium member of the All-Union Central Council of Trade Unions, a Central Committee member of the Communist Party, a member of the Supreme Soviet, a deputy people's commissar for foreign affairs and the head of the Soviet Information Bureau (Sovinformburo).

W25 : Scott 286-287 : Marshall Islands 1991 : USS Reuben James sunk, 1941.
Designer : David K Stone.
Selvedge commentary :
"The Navy is already at war in the Atlantic but the country doesn't seem to realise it"-
Admiral Harold R Stark, US navy.

The illustrations show *USS Reuben James* and a German *U-boat*.

USS *Reuben James* (DD-245) was a four-funnel *Clemson*-class destroyer that was
constructed after World War I. She was the first United States Navy ship to be
named after Boatswain's Mate Reuben James (1776–1838), who had
distinguished himself fighting in the First Barbary War, and was the first US ship
to be sunk by hostile action in the European Theatre of World War II.

Reuben James was laid down on 2 April 1919 by the New York Shipbuilding
Corporation of Camden, New Jersey, launched on 4 October 1919,
and commissioned on 24 September 1920. The destroyer was sunk by a torpedo
attack from German submarine *U-552* near Iceland on 31 October 1941, before
the United States had officially joined the war.

In August 1939, Harold Stark became Chief of Naval Operations with the rank
of admiral. In that position, he oversaw the expansion of the navy during 1940
and 1941, and its involvement in the Neutrality Patrols against
German submarines in the Atlantic during the latter part of 1941.

W26 : Scott 288- 291 : Marshall Islands 1991 : The Japanese attack Pearl Harbour, 1941.

Designer : David K Stone.

Selvedge Commentary :

"On December 7, 1941 — a date which will live in Famy — the United States of America was suddenly and deliberately attacked . . ." – Franklin D Roosevelt.

I fear that all we have done is to awaken a sleeping giant — and fill him with terrible resolve"- Admiral Isoroku Yamamoto.

The weapons are described as (1) American warplanes, (2) Japanese warplanes, (3) USS Arizona and (4) the Japanese aircraft carrier Akagi.

Pearl Harbour is an American lagoon harbour on the island of Oahu, Hawaii, west of Honolulu. It is also the headquarters of the United States Pacific Fleet. The U.S. government first obtained exclusive use of the inlet and the right to maintain a repair and coaling station for ships here in 1887. The surprise attack by the Imperial Japanese Navy on December 7, 1941, led the United States to declare war on the Empire of Japan, making the attack on Pearl Harbor the immediate cause of the United States' entry into World War II. Shortly after the attack, two American military commanders, Lt. Gen. Walter Short and Adm. Husband Kimmel, were demoted of their full ranks. The two American commanders later sought to restore their reputations and full ranks.

Isoroku Yamamoto (1884 – 1943) was a Marshal Admiral of the Imperial
Japanese Navy and the commander-in-chief of the Combined
Fleet during World War II. He commanded the fleet from 1939 until his death in
1943, overseeing the start of the Pacific War in 1941 and Japan's initial successes
and defeats before his plane was shot down by U.S. fighter aircraft over New
Guinea.

W27 : Scott 292 : Marshall Islands 1991 : The Japanese capture Guam, 1941.
Designer : Shannon Sternweis.
Selvedge commentary : *"I think the bittterest moment of my life came at sunrise when I saw the 'Rising Sun' ascend the flagpole . . . "* - Ensign Leona Jackson, Nurse Corps, US Navy.

Guam was an engagement during the Pacific War in World War II, and took place from 8 December to 10 December 1941 on Guam in the Mariana Islands between Japan and the United States. The American garrison was defeated by Japanese forces on 10 December, which resulted in an occupation until the Second Battle of Guam in 1944.

The Japanese landed about 400 troops of the 5th Defence Force from Saipan on Guam in the early morning of 10 December 1941 at Dungcas Beach, north of Agana. They attacked and quickly defeated the Insular Force Guard in Agana. They advanced on Piti, moving toward Sumay and the Marine barracks. The principal engagement took place on Agana's Plaza de España at 04:45 when a few Marines and Insular Force guardsmen fought with the Japanese naval soldiers. After token post-invasion resistance, the Marines, on McMillin's orders, surrendered at 05:45. McMillin officially surrendered at 06:00.[9] A few skirmishes took place all over the island before news of the surrender spread and the rest of the island forces laid down their arms. *YP-16* was scuttled by means of fire, and *YP-17* was captured by Japanese naval forces. An American freighter was damaged by the Japanese.

In the meantime the Japanese South Seas Detachment (about 5,500 men) under the command of Major General Tomitarō Horii made separate landings at Tumon Bay in the north, on the southwest coast near Merizo, and on the eastern shore of the island at Talofofo Bay.

U.S. Marine losses were five killed and 13 wounded (including the prior Japanese air assault of the island, the Marines' losses were 13 dead and 37 wounded). The U.S. Navy lost eight killed, and four of the Guam Insular Force Guards were killed and 22 others wounded. One Japanese naval soldier was killed and six wounded. Thirteen American civilians were killed by the Japanese during the battle. Six U.S. Navy seamen evaded capture by the Japanese rather than surrender; five were eventually retaken by the Japanese and beheaded, while Radioman First Class George Ray Tweed survived with the help of local Chamorros. They moved him from village to village, sometimes endangering their own families for his protection. The Japanese knew that an unknown American could not hide without some form of help. Consequently, Chamorro suspects were questioned, tortured, and beheaded. Despite the abuses, Chamorros loyal to the United States protected Tweed. Tweed managed to evade the Japanese during their occupation of Guam for 2 years and 7 months until he was rescued prior to the Second Battle of Guam in 1944.

Leona Jackson was appointed to the United States Navy Nurse Corps on 6 July 1936. She served her first few years, from 1936 until 1939 at the Naval Hospital, Philadelphia, Pennsylvania and then at the Naval Hospital, Brooklyn, New York from 1939 to 1940.[2]
In 1940, then-Ensign Jackson was assigned to the Naval Hospital, Guam, Marianas Islands. In December 1941, two days after Pearl Harbor, the Japanese invaded and took all personnel prisoner. Jackson and three other nurses, under the supervision of Chief Nurse Marian Olds, continued to work at the hospital until they were transported to Japan where they were held as prisoners of war until August 1942 when they were repatriated through Mozambique.

W28 : Scott 293 : Marshall Islands 1991 : The fall of Singapore to Japan, 1941. Designer : Brian Sanders.

Selvedge commentary :

"Unable . . . to continue to fight any longer. All ranks have done their best and are grateful for your help" - Lieutenant General Arthur Ernest Percival, February 18, 1941.

The fall of Singapore, also known as the Battle of Singapore, took place in the South–East Asian theatre of the Pacific War. The Japanese Empire captured the British stronghold of Singapore, with fighting lasting from 8 to 15 February 1942. Singapore was the foremost British military base and economic port in South–East Asia and had been of great importance to British interwar defence strategy. The capture of Singapore resulted in the largest British surrender in its history.

Before the battle, Japanese General Tomoyuki Yamashita had advanced with approximately 30,000 men down the Malayan Peninsula in the Malayan campaign. The British erroneously considered the jungle terrain impassable, leading to a swift Japanese advance as Allied defences were quickly outflanked. The British Lieutenant-General, Arthur Percival, commanded 85,000 Allied troops at Singapore, although many units were under-strength and most units lacked experience. The British outnumbered the Japanese but much of the water for the island was drawn from reservoirs on the mainland.

The British destroyed the causeway, forcing the Japanese into an improvised crossing of the Johore Strait. Singapore was considered so important that Prime Minister Winston Churchill ordered Percival to fight to the last man.

The Japanese attacked the weakest part of the island defences and established a beachhead on 8 February. Percival had expected a crossing in the north and failed to reinforce the defenders in time. Communication and leadership failures beset the Allies and there were few defensive positions or reserves near the beachhead. The Japanese advance continued and the Allies began to run out of supplies. By 15 February, about a million civilians in the city were crammed into the remaining area held by Allied forces, 1 percent of the island. Japanese aircraft continuously bombed the civilian water supply which was expected to fail within days. The Japanese were also almost at the end of their supplies and Yamashita wanted to avoid costly house-to-house fighting.

For the second time since the battle began, Yamashita demanded unconditional surrender and on the afternoon of 15 February, Percival capitulated. About 80,000 British, Indian, Australian and local troops became prisoners of war, joining the 50,000 taken in Malaya; many died of neglect, abuse or forced labour. Three days after the British surrender, the Japanese began the Sook Ching purge, killing thousands of civilians. The Japanese held Singapore until the end of the war.

Lieutenant-General Arthur Ernest Percival, (1887 – 1966) was a British Army officer. He saw service in the First World War and built a successful military career during the interwar period, but is best known for his defeat in the Second World War, when Percival commanded British Commonwealth forces during the Malayan campaign, which culminated in a catastrophic defeat at the Battle of Singapore.

Percival's surrender to the invading Imperial Japanese Army, which was the largest of its kind in British military history, significantly undermined Britain's prestige and military position in East Asia. Some historians, such as Sir John Smyth, have argued that under-funding of British Malaya's defences and the inexperienced, under-equipped nature of the Commonwealth forces in Malaya, not Percival's leadership, were ultimately to blame for the defeat.

Brigadier Sir John George Smyth, 1st Baronet, VC, MC, PC (1893 – 1983), often known as Jackie Smyth, was a British Indian Army officer and a Conservative Member of Parliament. During WWII, he led a unit in France and during the evacuation of Dunkirk, and in the Burma campaign.

W29 : Scott 294-295 : Marshall Islands 1991 : First combat by the Flying Tigers, 1941.

Designer : Brian Sanders.

Selvedge commentary :

"What Chennault did with his castoff ships and his hundred pilots is legend now. For six months they were thorns in the side of the Japanese Air Force" – General Robert Lee Scott, Jr.

The aircraft in the two images are the *American P-40* and the *Japanese* (near obsolete) *Ki-27 Fighter*.

In Asia, China and Japan had been at war since 1937.

China was already fighting its own civil war between the Nationalists of Chiang Kai-shek and Communist forces. The two sides came to a truce to fight against the Japanese. China, however, had little air power to fend off Japanese bombings.

The American Volunteer Group was largely the creation of Claire L. Chennault, a retired U.S. Army Air Corps officer who had worked in China since August 1937, first as military aviation advisor to Generalissimo Chiang Kai-shek in the early months of the Sino-Japanese War, then as director of a Chinese Air Force flight school centred in Kunming. Meanwhile, the Soviet Union supplied fighter and bomber squadrons to China, but these units were mostly withdrawn by the summer of 1940. Chiang then asked for American combat aircraft and pilots, sending Chennault to Washington as an adviser to China's ambassador and Chiang's brother-in-law, T. V. Soong.

Chennault spent the winter of 1940–1941 in Washington, supervising the purchase of 100 Curtiss P-40 fighters and the recruiting of 100 pilots and some 200 ground crew and administrative personnel that would constitute the 1st AVG. He also laid the groundwork for a follow-on bomber group and a second fighter group, though these would be aborted after the Pearl Harbor attack. 100 P-40 aircraft were crated and sent to Burma on third country freighters during spring 1941.

When Japanese aircraft attacked, Chennault's doctrine called for pilots to take on enemy aircraft in teams from an altitude advantage, since their aircraft were not as manoeuvrable or as numerous as the Japanese fighters they would encounter. He prohibited his pilots from entering into a turning fight with the nimble Japanese fighters, telling them to execute a diving or slashing attack and to dive away to set up for another attack. This "dive-and-zoom" technique was contrary to what the men had learned in U.S. service as well as what the Royal Air Force pilots in Burma had been taught; it had been used successfully, however, by Soviet units serving with the Chinese Air Force.

Robert Lee Scott Jr. (1908 – 2006) was a brigadier general in the United States Air Force and a flying ace of World War II, credited with shooting down 13 Japanese aircraft.
Scott is best known for his memoir, *God is My Co-Pilot* (1943), about his exploits in World War II with the Flying Tigers and the United States Army Air Forces in China and Burma.

W30 : Scott 296 : Marshall Islands 1991 : Fall of Wake Island to Japan, 1941.
Designer : David K Stone.
Selvedge commentary :
"The defence force had been simply overwhelmed by superior power in every quarter, on land, on sea and in the air"- Colonel Walter L J Bayler.

The aircraft shown are the Japanese *Mitsubishi G3M3 Medium bombers* opposed by the *US Grumman F4F Wildcats.*

The Battle of Wake Island was a battle of the Pacific campaign of World War II, fought on Wake Island. The assault began simultaneously with the attack on Pearl Harbor naval and air bases in Hawaii on the morning of 8 December 1941, (7 December in Hawaii), and ended on 23 December, with the surrender of American forces to the Empire of Japan. It was fought on and around the atoll formed by Wake Island and its minor islets of Peale and Wilkes Islands by the air, land, and naval forces of the Japanese Empire against those of the United States, with Marines playing a prominent role on both sides.

The battle started with a surprise bombing raid on December 8, 1941, within hours of Pearl Harbor, and the air raids continued almost every day for the duration of the battle. There were two amphibious assaults, one on December 11, 1941 (which was rebuffed) and another on December 23, that led to the Japanese capture of the atoll. In addition, there were several air battles above and around Wake and an encounter between two naval vessels. The U.S. lost control of the island and 12 fighter aircraft; in addition to the garrison being taken as prisoners of war, nearly 1200 civilian contractors were also captured by the Japanese. The Japanese lost about two dozen aircraft of different types, four surface vessels, and two submarines as part of the operation, in addition to at least 600 armed forces. It is typically noted that 98 civilian POWs captured in this battle were used for slave labour and then executed on Wake Island in October 1943. The other POWs were deported and sent to prisoner of war camps in Asia, with five executed on the sea voyage.

The island was held by the Japanese for the duration of the Pacific War theatre of World War II; the remaining Japanese garrison on the island surrendered to a detachment of United States Marines on 4 September 1945, after the earlier surrender on 2 September 1945 on the battleship USS *Missouri* in Tokyo Bay.

Walter Lewis John Bayler (1905 – 1984) was a brigadier general in the United States Marine Corps who was recognised during the war as the "Last Man Off Wake Island" and the only American to see combat at Wake Island, Midway and Guadalcanal. A naval aviator and communications engineer, he was at the forefront of the Marine Corps' use of radar for early warning and fighter direction. He was one of the driving forces behind the Marine Corps' establishment of an air warning program and served as the first commanding officer of the 1st Marine Air Warning Group (1st MAWG).

W31 : Scott 297 : Marshall Islands 1992 : The Arcadia Conference, Washington 1942.

Designer : Howard Koslow.

Selvedge commentary :

"We shall march forward together in comradeship until those who have sought to trample upon the rights of individual freedom . . . . are beat down . . . ." – Winston S Churchill, January 15, 1942.

Arcadia was the first meeting on military strategy between Britain and the United States; it came two weeks after the American entry into World War II. The Arcadia Conference was a secret agreement unlike the much wider postwar plans given to the public as the Atlantic Charter, agreed between Churchill and Roosevelt in August 1941.

From the start, significant differences in strategic priorities appeared. The British sought to push the Axis out of the Mediterranean, securing their lines of communications to their colonies. The American Navy, led by Admiral King, wished to prioritize fighting Japan, while the American Army, led by George C. Marshall, argued in favour of an immediate cross-channel invasion in 1942. Roosevelt, favouring naval strategy, was persuaded by Churchill to prioritize the Mediterranean, and even suggested to the Soviet Ambassador Litvinoff that a landing in North Africa might enable attacking German-occupied Europe from the south. Marshall, however, insisted upon a cross-channel invasion and suggested withdrawing from the liberation of Europe if the British did not agree to his plan. On Churchill's last day in Washington, the invasion of Guadalcanal was approved.

Roosevelt ultimately overruled Marshall after the British studied the feasibility of a cross-channel invasion and found it to be impossible in 1942. General Mark Clark, commander of all American forces in Britain, corroborated this conclusion later that year, pointing out that only one infantry division (the 34th Infantry Division) was available, but had neither amphibious training, anti-aircraft guns, tanks, nor landing craft. The 1st Armoured Division also lacked equipment, as were the new divisions arriving in-theatre.

The main policy achievements of Arcadia included the decision for "Germany First" (or "Europe first"—that is, the defeat of Germany was the highest priority); the establishment of the Combined Chiefs of Staff, based in Washington, for approving the military decisions of both the US and Britain; the principle of unity of command of each theatre under a supreme commander; drawing up measures to keep China in the war; limiting the reinforcements to be sent to the Pacific; and setting up a system for coordinating shipping. All the decisions were secret, except the conference drafted the Declaration by United Nations, which committed the Allies to make no separate peace with the enemy, and to employ full resources until victory.

In immediate tactical terms, the decisions at Arcadia included an invasion of North Africa in 1942, sending American bombers to bases in England, and for the British to strengthen their forces in the Pacific. Arcadia created a unified American-British-Dutch-Australian Command (ABDA) in the Far East; the ABDA fared poorly. It was also agreed at the conference to combine military resources under one command in the European Theatre of Operations.

W32 : Scott 298 : Marshall Islands 1992 : The fall of Manila to Japan, 1942.
Designer : Shannon Stirnweis.
Selvedge commentary :
"Manila is hereby declared an open city without the characteristics of a military objective" –
General Douglas MacArthur.

On Jan. 2, 1942, Japanese forces entered and occupied the city of Manila. The
following day, Gen. Masaharu Homma, the Japanese commander in chief,
announced the end of American colonial government and the imposition of
martial law. He also announced the establishment of the Japanese Military
Administration.

The Japanese launched the invasion by sea from Taiwan, over 200 miles
(320 km) north of the Philippines. The defending forces outnumbered the
Japanese by a ratio of 3:2 but were a mixed force of non-combat-experienced
regular, national guard, constabulary and newly created Commonwealth units.
The Japanese used first-line troops at the outset of the campaign, and by
concentrating their forces, they swiftly overran most of Luzon during the first
month.

The Japanese high command, believing that they had won the campaign, made a
strategic decision to advance by a month their timetable of operations in Borneo
and Indonesia and to withdraw their best division and the bulk of their airpower
in early January 1942. That, coupled with the defenders' decision to withdraw
into a defensive holding position in the Bataan Peninsula and also the defeat of
three Japanese battalions at the Battle of the Points and Battle of the Pockets,
enabled the Americans and Filipinos to hold out for four more months. After
the Japanese failure to penetrate the Bataan defensive perimeter in February, the
Japanese conducted a 40-day siege. The crucial large natural harbor and port
facilities of Manila Bay were denied to the Japanese until May 1942. While the
Dutch East Indies operations were unaffected, this heavily hindered the Japanese

offensive operations in New Guinea and the Solomon Islands, buying time for the U.S. Navy to make plans to engage the Japanese at Guadalcanal instead of much further east.

Japan's conquest of the Philippines is often considered the worst military defeat in U.S. history. About 23,000 American military personnel and about 100,000 Filipino soldiers were killed or captured.

W33 : Scott 299 : Marshall Islands 1992 : Capture of Raboul by Japan, 1942.
The commanders' images are shown; Admiral Isoroku Yamamoto and General
Douglas MacArthur, with their respective flags.
Designer : Chris Calle.
Selvedge commentary :
"*If the key to the Pacific for the Americans is the Marianas, the key for the Japanese is
Raboul*"- Admiral Ernest J King, Commander in Chief of the US Fleet.

Rabaul, the peacetime capital of the Australian Mandated Territory of New
Guinea, fell to the Japanese on 23 January 1942. The small Australian garrison,
Lark Force, was overwhelmed and most of its troops, including six army nurses,
captured. Approximately 400 of the troops escaped to the mainland and another
160 were massacred at Tol Plantation. In July 1942, about 1000 of the captured
Australian men, including civilian internees, were drowned when the Japanese
transport ship Montevideo Maru was sunk by an American submarine off the
Philippines coast en-route to Japan. Only the officers and nurses, sent to Japan
on a different ship, survived.

Rabaul was developed by the Japanese as their main naval base for the Solomon
Islands and New Guinea campaigns. Allied strategy for the South West Pacific
Area, Operation Cartwheel, aimed to isolate Rabaul and reduce it by air raids. In
December 1943, US Marines landed in New Britain and were soon replaced by
US Army troops who were relieved by the 5th Australian Division in late 1944.
The Australian established a base at Jacquinot Bay and in early 1945 cleared the
Japanese from the western end of the island pushing the Japanese into the
Gazelle peninsula with the 5th Division firmly established across the narrow
neck of the peninsula between Wide and Open Bays. This line remained quiet
until Japan surrendered in August 1945.

Isoroku Yamamoto (1884 –1943) was Marshal Admiral of the Imperial Japanese
Navy (IJN) and the commander-in-chief of the Combined Fleet during World
War II. Yamamoto held several important posts in the Imperial Navy, and

undertook many of its changes and reorganizations, especially its development of naval aviation. He was the commander-in-chief during the early years of the Pacific War and oversaw major engagements including the attack on Pearl Harbour and the Battle of Midway. Yamamoto was killed in April 1943 after American code breakers identified his flight plans, enabling the United States Army Air Forces to shoot down his aircraft.

Douglas MacArthur (1880 – 1964) was an American military leader who served as General of the Army for the United States. He served with distinction in WW1, and was Chief of Staff of the United States Army during the 1930s and played a prominent role in the Pacific Theatre during World War II. MacArthur was nominated for the Medal of Honour three times, and received it for his service in the Philippines campaign. He was one of only five men to rise to the rank of General of the Army in the U.S. Army, and the only one conferred the rank of field marshal in the Philippine Army. He had political ambitions and will appear again in the story.

W34 : Scott 300 : Marshall Islands 1992 : The Battle of the Java Sea, 1942.
Designer : Brian Sanders.
Selvedge commentary :
"The Dutch, British and American sailors fought to the last gun against impossible odds"-
Admiral Sir William James, Royal Navy, February 27, 1942.

Allied navies suffered a disastrous defeat at the hand of the Imperial Japanese
Navy on 27 February 1942 and in secondary actions over successive days.
The American-British-Dutch-Australian Command (ABDACOM) Strike Force
commander— Dutch Rear Admiral Karel Doorman—was killed. The aftermath
of the battle included several smaller actions around Java, including the smaller
but also significant Battle of Sunda Strait. These defeats led to Japanese
occupation of the entire Dutch East Indies.

The Japanese amphibious forces gathered to strike at Java, and on 27 February
1942 the main Allied naval force, under Rear Admiral Karel Doorman, sailed
northeast from Surabaya to intercept a convoy of the Japanese eastern invasion
force approaching from the Makassar Strait. The Japanese task force protecting
the convoy, commanded by Rear Admiral Takeo Takagi, consisted of two heavy
and two light cruisers and 14 destroyers.

The Allies eastern strike force, consisted of two heavy cruisers three light
cruisers, and nine destroyers

Karel Willem Frederik Marie Doorman (1889 – 1942) was a Dutch naval
officer commanded remnants of the short-lived American-British-Dutch-
Australian Command naval strike forces in the Battle of the Java Sea. He was
killed in action when his flagship HNLMS *De Ruyter* was torpedoed during the
battle.

Admiral Sir William Milbourne James, (1881 – 1973) was a British naval commander, politician and author. During the Second World War, James served as Commander-in-Chief, Portsmouth, from 1939. In 1940 he commanded Operation Aerial, the evacuation of British troops from Brittany and Normandy, a parallel operation to the Dunkirk evacuation.

Takeo Takagi (1892 – 1944) was an admiral in the Imperial Japanese Navy He was the commander of the IJN 6th Fleet, which oversaw the deployment of all submarines.

W35 : Scott 301 : Marshall Islands 1992 : The capture of Rangoon by Japan, 1942.

Designer : Brian Sanders.

Selvedge commentary :

"If Burma goes it seems to me our whole position, including that of Australia, will be in extreme peril" – Franklin D Roosevelt.

The recognised weapons of war include the *Japanese Type 3 Chi-Nu* armoured vehicle.

On 8 March 1942, Rangoon fell to the Japanese after two and a half months of heavy aerial bombing. At least 2,000 civilians were killed by the bombing, and another 400,000 became refugees. In a "scorched earth" policy, the city's dockyards, factories, oil refinery, and railways were all destroyed by the colonial authorities before they evacuated. Hundreds of offices and shops were looted by ordinary people. The fighting retreat from Rangoon (first to Mandalay and then to Assam) over the next three months would be the longest ever in British history. This "scorched earth" retreat and the subsequent battles to take back the country led to one of the most advanced countries in Southeast Asia becoming an economic ruin. The city was a crucial communication and industrial center in Burma and had the only port capable of handling troopships. Perhaps most importantly, strategically, the Burma Road began in Rangoon and allowed for a steady stream of military aid to be transported from Burma to Nationalist China. This supply route was essential for both Chiang Kai Shek's armies as well as allied forces in the region. As a result, the fall of Rangoon to the Japanese had significant consequences.

The Burma Road reopened in October 1940 and by late 1941 the U.S. was shipping munitions and other materials to supply the Chinese Army, whose continuing strength, in turn, forced the Japanese to keep considerable numbers of ground forces stationed in China. In fact, nearly half of the Imperial Army was stuck fighting Chiang Kai-shek's Nationalist forces. As a result, the Japanese decided it was necessary to close the Burma Road and cut off Chiang Kai Shek's lifeline. If successful, the Chinese would be able to free their forces for use elsewhere in the Pacific and perhaps gain complete control of China. Additionally, Burma was considered the gateway to gaining control of India. Overall, all parties involved in the Pacific War viewed the loss of Rangoon as the loss of Burma.

W36 : Scott 302 : Marshall Islands 1992 : Japanese land on New Guinea, 1942.
Designer David K Stone.

Selvedge commentary :

"Both armies had come to realise that whoever held this peninsular would command the northern approaches to Australia" – Historian William Manchester.

The New Guinea campaign of the Pacific War lasted from January 1942 until the end of the war in August 1945. During the initial phase in early 1942, the Empire of Japan invaded the Territory of New Guinea on 23 January and Territory of Papua on 21 July and overran western New Guinea (part of the Netherlands East Indies) beginning on 29 March. During the second phase, lasting from late 1942 until the Japanese surrender, the Allies—consisting primarily of Australian forces—cleared the Japanese first from Papua, then New Guinea, and finally from the Dutch colony.

The campaign resulted in a crushing defeat and heavy losses for the Empire of Japan. As in most Pacific War campaigns, disease and starvation claimed more Japanese lives than enemy action. Most Japanese troops never even encountered Allied forces and were instead simply cut off and subjected to an effective blockade by Allied naval forces. Garrisons were effectively besieged and denied shipments of food and medical supplies, and as a result some claim that 97% of Japanese deaths in this campaign were from non-combat causes. According to John Laffin, the campaign "was arguably the most arduous fought by any Allied troops during World War II."

William Raymond Manchester (1922 – 2004) was an American author, biographer, and historian. He served in the Battle of Okinawa, was severely wounded on June 5, 1945, and was promoted to sergeant[5] in July and awarded the Purple Heart.

On 6 March 1941 Australian author John Laffin enlisted in the Australian Imperial Force for service (boldly putting his age up by seven years). Three days after disembarking at Port Moresby on 4 December 1942, he was posted to the 2/31st Battalion, the history of which he would later write in Forever Forward (1994). (https://adb.anu.edu.au/biography/laffin-john).

W37 : Scott 303 : Marshall Islands 1992 : MacArthur evacuated from Corregidor, 1942.

Designer : Howard Koslow.

Selvedge commentary :

"I came through and I shall return"- General Douglas MacArthur.

During World War II, Corregidor was the site of two costly sieges and pitched battles—the first during the first months of 1942, and the second in February 1945—between the Imperial Japanese Army and the U.S. Army, along with its smaller subsidiary force, the Philippine Army.

During the Battle of the Philippines (1941–42), the Imperial Japanese Army invaded Luzon from the north (at Lingayen Gulf) in early 1942 and attacked Manila from its landward side. American and Filipino troops under the command of General Douglas MacArthur, retreated to the Bataan Peninsula, west of Manila Bay. The fall of Bataan on April 9, 1942, ended all organized opposition by the U.S. Armed Forces in the Far East (USAFFE) and gave way to the invading Japanese forces in Luzon in the northern Philippines. They were forced to surrender due to the lack of food and ammunition, leaving Corregidor and its adjacent islets at Manila Bay as the only areas in the region under U.S. control.

Between December 24, 1941, and February 19, 1942, Corregidor was the temporary location for the Government of the Philippines. On December 30, 1941, outside the Malinta Tunnel, Manuel L. Quezon and Sergio Osmeña were inaugurated respectively as president and vice-president of the Philippines Commonwealth for a second term.

General Douglas MacArthur also used Corregidor as Allied headquarters until March 11, 1942. The Voice of Freedom, the radio station of USAFFE broadcast from Corregidor, aired the infamous announcement of the fall of Bataan. In

April 1942, one battalion of the Fourth Marines was sent to reinforce the island's beach defences.

The Battle of Corregidor was the culmination of the Japanese campaign for the conquest of the Philippines. The fortifications across the entrance to Manila Bay were the remaining obstacle for the 14th Area Army of the Imperial Japanese Army led by Lieutenant General Masaharu Homma. American and Filipino soldiers on Corregidor and the neighbouring islets held out against the Japanese to deny the use of Manila Bay, but the Imperial Japanese Army brought heavy artillery to the southern end of Bataan, and proceeded north to blockade Corregidor. Japanese troops forced the surrender of the remaining American and Filipino forces on May 6, 1942, under the command of Lieutenant General Jonathan Wainwright.

The battle for the recapture of Corregidor occurred from February 16 to 26, 1945, in which American and Filipino forces successfully recaptured the island fortress from the Japanese occupying forces.

Jonathan Wainwright features on stamp # W42 later in the set.

W38 : Scott 304 : Marshall Islands 1992 : British raid on Saint Nazaire, 1942. Designer : Brian Sanders.

"*Stand by to ram*"- Lieutenant Commander S H Beattie, *HMS Campbelltown*.

The St Nazaire Raid or Operation Chariot was a British amphibious attack on the heavily defended Normandie dry dock at St Nazaire in German-occupied France during the Second World War. The operation was undertaken by the Royal Navy (RN) and British Commandos under the auspices of Combined Operations Headquarters on 28 March 1942. St Nazaire was targeted because the loss of its dry dock would force any large German warship in need of repairs, such as *Tirpitz*, sister ship of *Bismarck*, to return to home waters by running the gauntlet of the Home Fleet of the Royal Navy and other British forces, via the English Channel or the North Sea.

The obsolete destroyer HMS *Campbeltown*, accompanied by 18 smaller craft, crossed the English Channel to the Atlantic coast of France and rammed into the Normandie dry dock south gate. The ship had been packed with delayed-action explosives, well hidden within a steel and concrete case, that detonated later that day, putting the dock out of service until 1948.

A force of commandos landed to destroy machinery and other structures. German gunfire sank, set ablaze, or immobilized virtually all the small craft intended to transport the commandos back to England. The commandos fought their way through the town to escape overland but many surrendered when they ran out of ammunition or were surrounded by the Wehrmacht defending Saint-Nazaire.

Of the 612 men who undertook the raid, 228 returned to Britain, 169 were killed and 215 became prisoners of war. German casualties included over 360 dead, some of whom were killed after the raid when *Campbeltown* exploded. To recognise their bravery, 89 members of the raiding party were awarded decorations, including five Victoria Crosses. After the war, St Nazaire was one of 38 battle honours awarded to the commandos. The operation has been called "the greatest raid of all" in British military circles.

The then-Lieutenant Commander Stephen Beattie, a 33-year-old in command of the Royal Navy destroyer HMS *Campbeltown*, the tactical lynchpin of the St Nazaire Raid in 1942, was awarded the VC.

W39 : Scott 305 : Marshall Islands 1992 : Surrender of Bataan / Death March, 1942.

Designer : David K Stone.

Selvedge commentary :

"I made that march . . . in six days on one mess kit of rice. Other Americans made the 'March of Death' in twelve days without any food" - Colonel William F Dyass.

The Battle of Bataan; January 7 – April 9, 1942) was fought by the United States and the Philippine Commonwealth against Imperial Japan represented the most intense phase of the Japanese invasion of the Philippines. In January 1942, forces of the Imperial Japanese Army and Navy invaded Luzon along with several islands in the Philippine Archipelago after the bombing of the American naval base at Pearl Harbor.

The commander in chief of the U.S. and Filipino forces in the islands, General Douglas MacArthur, consolidated all of his Luzon-based units on the Bataan Peninsula to fight against the Japanese army. By this time, the Japanese controlled nearly all Southeast Asia. The Bataan Peninsula and the island of Corregidor were the only remaining Allied strongholds in the region.

Despite their lack of supplies, American and Filipino forces managed to fight the Japanese for three months, engaging them initially in a fighting retreat southward. As the combined American and Filipino forces made a last stand, the delay cost the Japanese valuable time and prevented immediate victory across the Pacific. The American surrender at Bataan to the Japanese, with 76,000 soldiers surrendering in the Philippines altogether, was the largest in American and Filipino military histories and was the largest United States surrender since the American Civil War's Battle of Harpers Ferry. Soon afterwards, U.S. and Filipino prisoners of war were forced into the Bataan Death March.

The continued resistance of the force on Bataan after Singapore and the Indies had fallen made heartening news among the Allied peoples. However, the extension of time gained by the defence was very largely a result of the transfer of the 48th Division from Homma's army at a critical time, and the exhaustion of the weakened force that remained. It cost a far stronger Japanese army as many days of actual combat to take Malaya and Singapore Island as it cost Homma to take Bataan and Corregidor.
The surrender of Bataan hastened the fall of Corregidor a month later. There is a suggestion that without the stand, the Japanese might have quickly overrun all the U.S. bases in the Pacific and could have quickly invaded Australia.

I can find no specific biographic detail for Colonel William F Dyass.

General George Marshall stated "These brutal reprisals upon helpless victims evidence the shallow advance from savagery which the Japanese people have made. ... We serve notice upon the Japanese military and political leaders as well as the Japanese people that the future of the Japanese race itself, depends entirely and irrevocably upon their capacity to progress beyond their aboriginal barbaric instincts".

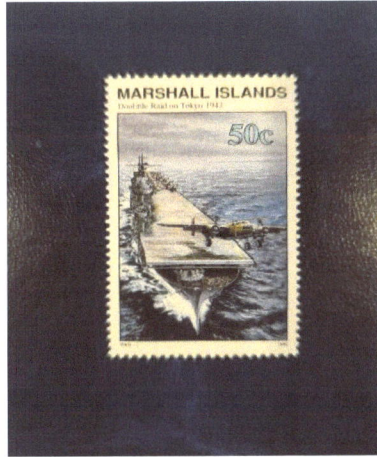

W40 : Scott 306 : Marshall Islands 1992 : The Doolittle Raid on Tokyo, 1942.
Designer David K Stone.
Selvedge commentary : *In my opinion, their flight was one of the most courageous deeds in all military history*"- Admiral W F Halsey.

The recognised weapons of war are the aircraft carrier *USS Hornet* and a *Boeing B25 bomber*.

The Doolittle Raid, was an air raid on 18 April 1942 by the United States on the Japanese capital Tokyo and other places on Honshu during the war. It was the first American air operation to strike the Japanese archipelago. Although the raid caused comparatively minor damage, it demonstrated that the Japanese mainland was vulnerable to American air attacks. It served as an initial retaliation for the December 7, 1941, attack on Pearl Harbor, and provided an important boost to American morale. The raid was named after Lieutenant Colonel James Doolittle, who planned and led the attack. It was one of six American carrier raids against Japan and Japanese-held territories conducted in the first half of 1942.

Under the final plan, 16 B-25B Mitchell medium bombers, each with a crew of five, were launched from the US Navy aircraft carrier USS *Hornet*, in the Pacific Ocean. There were no fighter escorts. After bombing the military and industrial targets, the crews were to continue westward to land in China.

On the ground, the raid killed about 50 people and injured 400. Damage to Japanese military and industrial targets was slight, but the raid had major psychological effects. In the United States, it raised morale. In Japan, it raised fear and doubt about the ability of military leaders to defend the home islands, but the bombing and strafing of civilians created a desire for retribution—this was exploited for propaganda purposes. The raid also pushed forward Admiral Isoroku Yamamoto's plans to attack Midway Island in the Central Pacific—an attack that turned into a decisive defeat of the Imperial Japanese Navy (IJN) by the US Navy in the Battle of Midway. The consequences of the Doolittle Raid were most severely felt in China: in reprisal for the raid, the Japanese launched the Zhejiang-Jiangxi campaign, killing 250,000 civilians and 70,000 soldiers.

William Frederick "Bull" Halsey Jr. (1882 – 1959) was a Navy admiral . He is one of four officers to have attained the rank of five-star fleet admiral of the United States Navy. Halsey was made commander of the South Pacific Area, and led the Allied forces over the course of the Battle for Guadalcanal (1942–1943) and the fighting up the Solomon chain (1942–1945). In 1943 he was made commander of the Third Fleet, the post he held through the rest of the war. He took part in the Battle of Leyte Gulf, the largest naval battle of the Second World War and, by some criteria, the largest naval battle in history.

W41 : Scott 307 : Marshall Islands 1942 : The fall of Corregidor, Philippines 1942.

Selvedge commentary :

"With broken heart and head bowed in sadness . . . I must arrange terms for the surrender" - Major General Jonathon Wainwright to President Franklin D Roosevelt. May 6, 1942.

General Wainwright commanded American and Filipino forces during the Japanese invasion of the Philippines, for which he received a Medal of Honour for his courageous leadership. In May 1942, on the island stronghold of Corregidor, lacking food, supplies and ammunition, in the interest of minimizing casualties Wainwright surrendered the remaining Allied forces on the Philippines. In August 1945, he was rescued by the Red Army in Manchukuo and participated in the formal Japanese Surrender.

W42 : Scott 308-311 : Marshall Islands 1992 : The Battle of the Coral Sea, 1942.
Designer : Brian Sanders.
Selvedge commentary :
"Remember the folks back home are counting on us"- Navy Lieutenant John Powers, recipient of the Medal of Honour.
" . . . a victory with decisive and far reaching consequences" – Admiral Chester Nimitz, Commander in Chief, US Pacific Fleet.

The recognised weapons are (1) *USS Lexington*, (2) *Japanese Mitsubishi A6M2 'Zeros'*, (3) *US Douglas SBD Dauntless dive-bombers*, (4) *Japanese carrier Shōkaku*

I was aware that this sheet / block of four images was reissued. Following the advice of Brookman Stamp Company, who suggested that the reissues were because of "errors", I have spent time considering the image above and the image descriptions. Although I have not been able to acquire a copy of the reissued image #(3) I believe this is the error. The aircraft are torpedo bombers not dive bombers. The description should read US Douglas TDB Devastator not SBD dive-bombers.

Wikipedia confirms the TDB Devastator 'role in the Coral Sea Battle :
"In the early days of the Pacific war, the TBD acquitted itself well during February and March 1942, with TBDs

from *Enterprise* and *Yorktown* attacking targets in the Marshall and Gilbert Islands, Wake and Marcus Islands, while TBDs from *Yorktown* and *Lexington* struck Japanese shipping off New Guinea on 10 March.[15] In the Battle of the Coral Sea Devastators helped sink the Japanese aircraft carrier *Shōhō* on 7 May, but failed to hit another carrier, the *Shōkaku* the next day" (en.wikipedia.org/wiki/Douglas_TBD_Devastator).

The Battle of the Coral Sea, from 4 to 8 May 1942, was a major naval battle between the Imperial Japanese Navy (IJN) and naval and air forces of the United States and Australia. Taking place in the Pacific Theatre, the battle was the first naval action in which the opposing fleets neither sighted nor fired upon one another, attacking over the horizon from aircraft carriers instead.

USS *Lexington* was the name ship of her class of two aircraft carriers built for the United States Navy during the 1920s. Originally designed as a battlecruiser, she was converted into one of the Navy's first aircraft carriers during construction to comply with the terms of the Washington Naval Treaty of 1922, which essentially terminated all new battleship and battlecruiser construction. The ship entered service in 1928 and was assigned to the Pacific Fleet for her entire career. *Lexington* and her sister ship, *Saratoga*, were used to develop and refine carrier tactics in a series of annual exercises before World War II.

Shōkaku was the lead ship of her class of two aircraft carriers built for the Imperial Japanese Navy (IJN) shortly before the Pacific War. Along with her sister ship *Zuikaku*, she took part in several key naval battles during the war, including the attack on Pearl Harbor, the Battle of the Coral Sea, and the Battle of the Santa Cruz Islands, before being torpedoed and sunk by the U.S. submarine USS *Cavalla* at the Battle of the Philippine Sea.

As the fleet moved to prevent further Japanese expansion in the Solomons, pilot John Powers took part in the May 4, 1942, raid on Tulagi, flying without fighter cover to score two hits on Japanese ships. As the main Battle of the Coral Sea developed on May 7, 1942, Powers and his companions discovered carrier *Shōhō* and, bombing at extremely low altitudes, sank her in 10 minutes. The next morning, May 8, while the carrier battle continued, he joined the attack on the carrier *Shokaku*, scoring an important bomb hit. Powers' low-bombing run brought him into heavy antiaircraft fire, and his plane plunged into the sea.

Powers was declared dead and, for his actions in this series of attacks, he was posthumously awarded the Medal of Honour.

On 24 March 1942, the newly formed US-British Combined Chiefs of Staff issued a directive designating the Pacific theatre an area of American strategic responsibility. Six days later, the US Joint Chiefs of Staff (JCS) divided the theatre into three areas: the Pacific Ocean Areas, the Southwest Pacific Area (commanded by General Douglas MacArthur), and the Southeast Pacific Area. The JCS designated Nimitz as "Commander in Chief, Pacific Ocean Areas", with operational control over all Allied units (air, land, and sea) in that area.

W43 : Scott 312-315 : Marshall Islands 1992 : The Battle of Midway 1942.
Selvedge Commentary :
"*Midway was one of the Decisive Naval Battles of all history*"– Masanori Ito (The Battle for Japan 1944-45, p47).
"*A momentous victory in the making*"- Admiral Chester Nimitz, Fleet Admiral US Navy.

The recognised weapons are (1) The *USS Lexington* and *US Grumman F4F-3 Wildcats*, (2) *Japanese ship Aichi* and *Natajima Kate aircraft*, (3) *US Douglas SPD-3 Devastators* and (4) the Japanese carrier *Shōhō* and *Mitsubishi A6M Zeros*.

The Battle of Midway was a major naval battle in the Pacific Theater of World War II that took place 4–7 June 1942, six months after Japan's attack on Pearl Harbor and one month after the Battle of the Coral Sea. The U.S. Navy under Admirals Chester W. Nimitz, Frank J. Fletcher, and Raymond A.
Spruance defeated an attacking fleet of the Imperial Japanese Navy under Admirals Isoroku Yamamoto, Chūichi Nagumo, and Nobutake Kondō north of Midway Atoll, inflicting devastating damage on the Japanese fleet. Military historian John Keegan called it "the most stunning and decisive blow in the history of naval warfare".

The Battle of Midway has often been called "the turning point of the Pacific". It was the Allies' first major naval victory against the Japanese. Had Japan won the battle as thoroughly as the U.S. did, it might have been able to capture Midway Island. *Saratoga* would have been the only American carrier in the Pacific, as no new ones were completed before the end of 1942. While the U.S. would probably not have sought peace with Japan as Yamamoto hoped, his country might have revived Operation FS to invade and occupy Fiji and Samoa; attacked Australia, Alaska, and Ceylon; or even attempted to occupy Hawaii.

Masanori Ito was a Japanese journalist, author, and one of Japan's leading military commentators. He personally knew many of the Japanese naval commanders. He was the author of the book The End of the Imperial Japanese Navy), first printed in 1956, which was a dramatic account offering a rare glimpse into Japanese naval warfare in the Pacific.

Chester William Nimitz (1885 – 1966) was a fleet admiral in the United States Navy. He played a major role in the as Commander in Chief, US Pacific Fleet, and Commander in Chief, Pacific Ocean Areas, commanding Allied air, land, and sea forces.

W 44 : Scott 316 : Marshall Islands 1992 : The destruction of Lidice,
Czechoslovakian Village, 1942.

Designer : Howard Koslow.

Selvedge commentary :

"*Remember the murder of Lidice! And pray for your mother and me*"- From 'The Murder
of Lidice' by Edna St. Vincent Millay, 1942.

On June 10, 1942, Nazi troops obliterate the village of Lidice, Czechoslovakia
after killing all adult males and deporting most of the surviving women and
children to concentration camps. The brutal action came as part of a retaliation
for the assassination of Nazi SS leader Reinhard Heydrich, known as *Heydrich the
Hangman* - despite there being no solid evidence connecting the town to the
assassination plot.

The massacre was carried out a day after the Nazis rounded up the residents of
Lidice, located near Prague. SS troops herded all the town's male residents aged
16 and older—more than 170—to a local farmstead and gunned them down.
Germans shot seven women trying to flee, and deported the remaining women
to Ravensbruck concentration camp, where about 50 died and three were
recorded as "disappeared." Of some 105 children in the village, one was reported
shot while running away, approximately 80 were reported murdered in Chelmno
killing centre, and a handful were reported murdered in German Lebensborn
orphanages. A few of the orphans, deemed "racially pure" by Nazi standards,
were dispersed throughout German territory to be renamed and raised as
Germans.

Reinhard Tristan Eugen Heydrich (1904 – 1942) was a high-ranking
German SS and police official during the Nazi era and a principal architect of the
Holocaust.
Heydrich was chief of the Reich Security Main Office (including
the Gestapo, Kripo, and SD). He was also *Stellvertretender
Reichsprotektor* (Deputy/Acting Reich-Protector) of Bohemia and Moravia. He
chaired the January 1942 Wannsee Conference which formalised plans for the

"Final Solution to the Jewish question"—the deportation and genocide of all Jews in German-occupied Europe.

Edna St. Vincent Millay (1892 – 1950) was an American lyrical poet and playwright.
In 1942 in *The New York Times Magazine*, Millay mourned the destruction of the Czech village Lidice. Millay wrote: "The whole world holds in its arms today / The murdered village of Lidice, / Like the murdered body of a little child."

W45 : Scott 317 : Marshall Islands 1992 : Sevastopol falls to the Germans, 1942.
Designer : Brian Sanders.
Selvedge commentary : *"We shall return . . . the lamps of the Chersonese Lighthouse went out, but we shall light them again"*- An anonymous Russian soldier".

The recognised gun is a Germany's Schwerer Gustav railway weapon.

As the outbreak of World War II approached, Nazi Germany ordered artillery manufacturers Krupp and Rheimetall-Borsig to build several super-heavy siege guns, vital to smash through French and Belgian fortresses that stood in the way of the Blitzkrieg. These 'secret weapons' were much larger than the siege artillery of World War I and included the largest artillery piece of the war, the massive 80cm railway gun 'schwere Gustav' (Heavy Gustav). However, these complex and massive artillery pieces required years to build and test and, as war drew near, the German High Command hastily brought several WWI-era heavy artillery pieces back into service and then purchased, and later confiscated, a large number of Czech Skoda mortars.

The new super siege guns began entering service in time for the invasion of Russia, notably participating in the attack on the fortress of Brest-Litovsk. The highpoint for the siege artillery was the siege of Sevastopol in the summer of 1942, which saw the largest concentration of siege guns in the war. Afterwards, when Germany was on the defensive in the second half of 1943, the utility of the guns was greatly diminished, and they were employed in a piecemeal and sporadic fashion on both the Eastern and Western Fronts. In total, the German Army used some 50 siege guns during World War II, far more than the thirty-five it had during World War I.

W46 : Scott 318-319 : Marshall Islands 1992 : Convoy PQ-17 destroyed, 1942.
Designer : Brian Sanders.
Selvedge commentary :
"... *convoy is to disperse"and proceed to Russian ports. Convoy is to scatter"*- Admiralty to
Rear Admiral L H K Hamilton, July 4, 1942.
"One of the most melancholy naval episodes in the whole of the war" – Winston S
Churchill.

The two images are simply titled *British Merchant ship* and *German U-boat*.

Convoy PQ 17 was the code name for an Allied Arctic convoy . On 27 June
1942, the ships sailed from Hvalfjörður, Iceland, for the port of Arkhangelsk in
the Soviet Union. The convoy was located by German forces on 1 July, after
which it was shadowed continuously and attacked. The First Sea
Lord Admiral Dudley Pound, acting on information that German ships,
including the German battleship *Tirpitz*, were moving to intercept, ordered
the covering force, based on the Allied battleships HMS *Duke of
York* and USS *Washington* away from the convoy and told the convoy to scatter.
Because of vacillation by *Oberkommando der Wehrmacht* (OKW, German armed
forces high command), the *Tirpitz* raid never materialised. The convoy was the
first large joint Anglo-American naval operation under British command; in
Churchill's view this encouraged a more careful approach to fleet movements.

As the close escort and the covering cruiser forces withdrew westwards to intercept the German raiders, the merchant ships were left without escorts. The merchant ships were attacked by *Luftwaffe* aircraft and U-boats and of the 35 ships, only eleven reached their destination, delivering 70,000 long tons (71,000 metric tons) of cargo. The convoy disaster demonstrated the difficulty of passing adequate supplies through the Arctic, especially during the summer, with the midnight sun. The German success was possible through German signals intelligence and cryptological analysis.

W47 : Scott 320 : Marshall Islands 1992 : Marines land at Guadalcanal, 1942.
Designer : Howard Koslow.
Selvedge commentary :
"For us who were there, Guadacanal is not a name but an emotion . . ." Historian Samuel
Eliot Morison.

On 7 August 1942, Allied forces, predominantly United States Marines, landed
on Guadalcanal, Tulagi, and Florida Island in the southern Solomon Islands,
with the objective of using Guadalcanal and Tulagi as bases in supporting a
campaign to eventually capture or neutralize the major Japanese base
at Rabaul on New Britain. The Japanese defenders, who had occupied those
islands since May 1942, were outnumbered and overwhelmed by the Allies, who
captured Tulagi and Florida, as well as the airfield—later named Henderson
Field—that was under construction on Guadalcanal.

Surprised by the Allied offensive, the Japanese made several attempts between
August and November to retake Henderson Field. Three major land battles,
seven large naval battles (five nighttime surface actions and two carrier battles),
and almost daily aerial battles culminated in the decisive Naval Battle of
Guadalcanal in early November, with the defeat of the last Japanese attempt to
bombard Henderson Field from the sea and to land enough troops to retake it.
In December, the Japanese abandoned their efforts to retake Guadalcanal
and evacuated their remaining forces by 7 February 1943, in the face of an
offensive by the U.S. Army's XIV Corps, with the Battle of Rennell Island, the
last major naval engagement, serving to secure protection for the Japanese
troops to evacuate safely.
The campaign followed the successful Allied defensive actions at the Battle of
the Coral Sea and the Battle of Midway in May and June 1942. Along with the
battles at Milne Bay and Buna–Gona, the Guadalcanal campaign marked the
Allies' transition from defensive operations to offensive ones and effectively
allowed them to seize the strategic initiative in the Pacific theatre from the
Japanese. The campaign was followed by other Allied offensives in the Pacific,

most notably: the Solomon Islands campaign, New Guinea campaign, the Gilbert and Marshall Islands campaign, the Mariana and Palau Islands campaign, the Philippines campaign (1944–1945), and the Volcano and Ryukyu Islands campaign prior to the surrender of Japan in August, 1945.

Samuel Eliot Morison (1887 – 1976) was an American historian noted for his works of maritime history and American history that were both authoritative and popular.

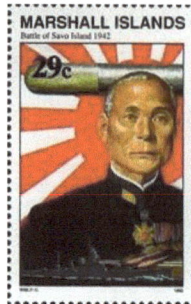

W48 : Scott 321 : Marshall Islands 1992 : The Battle of Savo Island, 1942.
Designer : Shannon Stirnweis.
Selvedge commentary :
"The Japanese did not take advantage . . . because they did not know how severe our losses were" - Admiral Ernest J King, Commander, US Fleet.

The Battle of Savo Island, was the naval battle of the Solomon Islands campaign of the Pacific War of World War II between the Imperial Japanese Navy and Allied naval forces. The battle took place on 8–9 August 1942 and was the first major naval engagement of the Guadalcanal campaign and the first of several naval battles in the straits, near the island of Guadalcanal.

The Imperial Japanese Navy, in response to Allied amphibious landings in the eastern Solomon Islands, mobilized a task force of seven cruisers and one destroyer under the command of Vice Admiral Gunichi Mikawa. In a night action, Mikawa thoroughly surprised and routed the Allied force, sinking one Australian and three American cruisers, while suffering only light damage in return.

After the initial engagement, Mikawa, fearing Allied carrier strikes against his fleet in daylight, decided to withdraw under cover of night rather than attempt to locate and destroy the Allied invasion transports. The Japanese attacks prompted the remaining Allied warships and the amphibious force to withdraw earlier than planned (before unloading all supplies), temporarily ceding control of the seas around Guadalcanal to the Japanese. This early withdrawal of the fleet left the Allied ground forces (primarily United States Marines), which had landed on Guadalcanal and nearby islands only two days before, in a precarious situation with limited supplies, equipment, and food to hold their beachhead.

The battle was the first of five costly, large-scale sea and air-sea actions fought in support of the ground battles on Guadalcanal, as the Japanese sought to counter the American offensive in the Pacific. These sea battles took place after increasing delays by each side to regroup and refit, until the 30 November 1942 Battle of Tassafaronga—after which the Japanese, eschewing the costly losses, attempted resupplying by submarine and barges.

The final naval battle, the Battle of Rennell Island, took place months later on 29–30 January 1943, by which time the Japanese were preparing to evacuate their remaining land forces and withdraw.

Gunichi Mikawa (1888 – 1981) was the commander of a heavy cruiser force that defeated the U.S. Navy and the Royal Australian Navy at the Battle of Savo Island in Ironbottom Sound on the night of 8–9 August 1942.
In this battle, his squadron of cruisers, plus one destroyer, sank three USN cruisers, plus the RAN heavy cruiser HMAS *Canberra*; Mikawa's force suffered no losses in the actual battle, although heavy cruiser *Kako* was sunk by the undetected American submarine *S-44* on the return to their base near Rabaul in the Bismarck Archipelago.

Ernest Joseph King (1878 – 1956) was a fleet admiral in the United States Navy who served as Commander in Chief, United States Fleet (COMINCH) and Chief of Naval Operations (CNO) during World War II. Franklin Delano Roosevelt appointed King to command global American strategy during World War II and he held supreme naval command in his unprecedented double capacity as COMINCH and CNO.

OFFICIAL FIRST DAY COVER

MARSHALL ISLANDS

Battle of Savo Island
August 9, 1942

W49 : Scott 322 : Marshall Islands 1992 : The Dieppe Raid, 1942.
Designer : Brian Sanders.
Selvedge Commentary :
"For every soldier who died at Dieppe ten were saved on D-Day"- Lord Louis
Mountbatten.

Operation Jubilee or the Dieppe Raid (19 August 1942) was a
disastrous Allied amphibious attack on the German-occupied port of Dieppe in
northern France, during the Second World War. Over 6,050 infantry,
predominantly Canadian, supported by a regiment of tanks, were put ashore
from a naval force operating under the protection of Royal Air Force (RAF)
fighters.

The port was to be captured and held for a short period, to test the feasibility of
a landing and to gather intelligence. German coastal defences, port structures
and important buildings were to be demolished. The raid was intended to boost
Allied morale, demonstrate the commitment of the United Kingdom to re-open
the Western Front and support the Soviet Union, fighting on the Eastern Front.
The *Luftwaffe* made a maximum effort against the landing as the RAF had
expected, and the RAF lost 106 aircraft (at least 32 to anti-aircraft fire or
accidents) against 48 German losses. The Royal Navy lost 33 landing craft and a
destroyer. Aerial and naval support was insufficient to enable the ground forces
to achieve their objectives. The tanks were trapped on the beach and the infantry
was largely prevented from entering the town by obstacles and German fire.[7]

After less than six hours, mounting casualties forced a retreat. Within ten hours,
3,623 of the 6,086 men who landed had been killed, wounded, or taken prisoner.
5,000 were Canadians, who suffered a 68% casualty rate, with 3,367 killed,
wounded or taken prisoner. The operation was a fiasco in which only one
landing force temporarily achieved its objective, and a small amount of military
intelligence was gathered.

Both sides learnt important lessons regarding coastal assaults. The Allies learnt lessons that influenced the success of the D-Day landings. Artificial harbours were declared crucial, tanks were adapted specifically for beaches, a new integrated tactical air force strengthened ground support, and capturing a major port at the outset was no longer seen as a priority. Churchill and Mountbatten both claimed that these lessons had outweighed the cost. The Germans also believed that Dieppe was a learning experience and made a considerable effort to improve the way they defended the occupied coastlines of Europe.

Admiral of the Fleet Louis Mountbatten, 1st Earl Mountbatten of Burma (born Prince Louis of Battenberg; 1900 – 1979), commonly known as Lord Mountbatten, was a British statesman, Royal Navy officer and close relative of the British royal family. He was born in the United Kingdom to the prominent Battenberg family. He was a maternal uncle of Prince Philip, Duke of Edinburgh, and a second cousin of King George VI. He joined the Royal Navy during the First World War and was appointed Supreme Allied Commander, South East Asia Command, in the Second World War. He later served as the last Viceroy of India and briefly as the first Governor-General of the Dominion of India.

W50 : Scott 323 : Marshall Islands 1992 : The Battle of Stalingrad, 1942.
Designer : David K Stone.
Selvedge commentary :
"*We shall hold the city or die there*"- General Vasily Chuikov.

The Battle of Stalingrad (17 July 1942 – 2 February 1943) was the major battle
on the Eastern Front of World War II, beginning when Nazi Germany and
its Axis allies attacked and became locked in a protracted struggle with the Soviet
Union for control over the Soviet city of Stalingrad in southern Russia. The
battle was characterized by fierce close-quarters combat and direct assaults on
civilians in aerial raids; the battle epitomized urban warfare, being the single
largest and costliest urban battle in military history. It was the bloodiest and
fiercest battle of the entirety of World War II—and arguably in all of human
history—as both sides suffered tremendous casualties amidst ferocious fighting
in and around the city. Today, the Battle of Stalingrad is commonly regarded as
the turning point in the European theatre of World War II, as
Germany's *Oberkommando der Wehrmacht* was forced to withdraw a considerable
amount of military forces from other regions to replace losses on the Eastern
Front. By the time the hostilities ended, the German 6th Army and 4th Panzer
Army had been destroyed and Army Group B was routed. The Soviets' victory at
Stalingrad shifted the Eastern Front's balance of power in their favour, while also
boosting the morale of the Red Army.

Both sides placed great strategic importance on Stalingrad, for it was the largest
industrial centre of the Soviet Union and an important transport hub on
the Volga River: controlling Stalingrad meant gaining access to the oil fields of

the Caucasus and having supreme authority over the Volga River. As the conflict progressed, Germany's fuel supplies dwindled and thus drove it to focus on moving deeper into Soviet territory and taking the country's oil fields at any cost. The German military first clashed with the Red Army's Stalingrad Front on the distant approaches to Stalingrad on 17 July. On 23 August, the 6th Army and elements of the 4th Panzer Army launched their offensive with support from intensive bombing raids by the *Luftwaffe*, which reduced much of the city to rubble. The battle soon degenerated into house-to-house fighting, which escalated drastically as both sides continued pouring reinforcements into the city. By mid-November, the Germans, at great cost, had pushed the Soviet defenders back into narrow zones along the Volga's west bank. However, winter set in within a few months and conditions became particularly brutal, with temperatures often dropping tens of degrees below freezing. In addition to fierce urban combat, brutal trench warfare was prevalent at Stalingrad as well.

On 19 November, the Red Army launched Operation Uranus, a two-pronged attack targeting the Romanian armies protecting the 6th Army's flanks. The Axis flanks were overrun, and the 6th Army was encircled. Adolf Hitler was determined to hold the city for Germany at all costs and forbade the 6th Army from trying a breakout; instead, attempts were made to supply it by air and to break the encirclement from the outside. Though the Soviets were successful in preventing the Germans from making enough airdrops to the trapped Axis armies at Stalingrad, heavy fighting continued for another two months. On 2 February 1943, the 6th Army, having exhausted their ammunition and food, finally capitulated after several months of battle, making it the first of Hitler's field armies to have surrendered.

Vasiky Chuikov, was the commander the Russian 62nd Army which saw heavy combat during the Battle of Stalingrad.

W 51 : Scott 324 : Marshall Islands 1992 : The Battle of the Eastern Solomon Islands, 1942.

Designer : David K Stone.

Selvedge commentary :

"The Japanese . . . with air forces seriously reduced were retiring" - Admiral Chester Nimitz.

The naval Battle of the Eastern Solomons (also known as the Battle of the Stewart Islands and in Japanese sources as the Second Battle of the Solomon Sea) took place on 24–25 August 1942 and was the third carrier battle of the Pacific campaign of World War II and the second major engagement fought between the United States Navy and the Imperial Japanese Navy during the Guadalcanal campaign. As at the Battle of the Coral Sea and the Battle of Midway, the ships of the two adversaries were never within sight of each other. Instead, all attacks were carried out by carrier-based or land-based aircraft.

After several damaging air attacks, the naval surface combatants from both America and Japan withdrew from the battle area. Although neither side secured a clear victory, the U.S. and its allies gained a tactical and strategic advantage. Japan's losses were greater and included dozens of aircraft and their experienced aircrews. Also, Japanese reinforcements intended for Guadalcanal were delayed and eventually delivered by warships rather than transport ships, giving the Allies more time to prepare for the Japanese counteroffensive and preventing the Japanese from landing heavy artillery, ammunition, and other supplies.

W52 : Scott 325 : Marshall Islands 1992 : The Battle of Cape Esperance, 1942.
Designer : David K Stone.
Selvedge commentary :
"We . . . turned back the Japanese . . . and shattered their carrier air strength on the eve of the critical days of mid-November" - Admiral Chester Nimitz.

The Battle of Cape Esperance, also known as the Second Battle of Savo Island , took place on 11–12 October 1942, in the Pacific campaign of World War II between the Imperial Japanese Navy and United States Navy. The naval battle was the second of four major surface engagements during the Guadalcanal campaign and took place at the entrance to the strait between Savo Island and Guadalcanal in the Solomon Islands. Cape Esperance (9°15'S 159°42'E) is the northernmost point on Guadalcanal, and the battle took its name from this point.

On the night of 11 October, Japanese naval forces in the Solomon Islands area—under the command of Vice Admiral Gunichi Mikawa—sent a major supply and reinforcement convoy to their forces on Guadalcanal. The convoy consisted of two seaplane tenders and six destroyers and was commanded by Rear Admiral Takatsugu Jojima. At the same time but in a separate operation, three heavy cruisers and two destroyers—under the command of Rear Admiral Aritomo Gotō were to bombard the Allied airfield on Guadalcanal (called Henderson Field by the Allies) with the object of destroying Allied aircraft and the airfield's facilities.

Shortly before midnight on 11 October, a U.S. force of four cruisers and five destroyers—under the command of Rear Admiral Norman Scott—intercepted Gotō's force as it approached Savo Island near Guadalcanal. Taking the Japanese by surprise, Scott's warships sank one of Gotō's cruisers and one of his destroyers, heavily damaged another cruiser, mortally wounded Gotō, and forced the rest of Gotō's warships to abandon the bombardment mission and retreat. During the exchange of gunfire, one of Scott's destroyers was sunk, and one cruiser and another destroyer were heavily damaged. In the meantime, the Japanese supply convoy successfully completed unloading at Guadalcanal and began its return journey without being discovered by Scott's force. Later on the morning of 12 October, four Japanese destroyers from the supply convoy turned back to assist Gotō's retreating, damaged warships. Air attacks by U.S. aircraft from Henderson Field sank two of these destroyers later that day.

As with the preceding naval engagements around Guadalcanal and to be expected from a battle of relatively limited size, the strategic outcome was inconclusive because neither the Japanese nor United States navies secured operational control of the waters around Guadalcanal as a result of this action, and a heavy bombardment operation against Henderson Field would be conducted soon after, causing severe destruction in the three nights between 13 and 16 October. However, the Battle of Cape Esperance provided a significant morale boost to the U.S. Navy after its disastrous defeat at the Battle of Savo Island.

Norman Scott (1889 – 1942) was killed along with many of his staff when the ship he was on – the light cruiser USS *Atlanta* – was hit by gunfire from the heavy cruiser USS *San Francisco* during the nighttime fighting in the Naval Battle of Guadalcanal.

Admiral Aritomo Gotō (1888 – 1942), on 10 September 1941 he was placed in command of Cruiser Division 6 (CruDiv6). On 11 October, the remaining three cruisers of CruDiv6 approached Guadalcanal at night to bombard the Allied airbase at Henderson Field as well as to support a large "Tokyo Express" run occurring the same evening. Gotō's force was surprised by a force of American cruisers and destroyers under the command of U.S. Rear Admiral Norman Scott. In the resulting Battle of Cape Esperance, Gotō was mortally wounded onboard *Aoba* and died later.

Takatsugu Jōjima (1890 – 1967) was an admiral in the Imperial Japanese Navy. Jōjima was captain of *Shōkaku*during the attack on Pearl Harbor, the Battle of Rabaul, the Indian Ocean Raid, Operation Mo, the Battle of the Coral Sea. Promoted to rear admiral on 1 May 1942. He was commander of naval aviation units throughout the war, he also led the seaplane tenders of R-Area Air Force that participated in the defence of Guadalcanal during the Guadalcanal campaign including the Battle of Cape Esperance and Japanese efforts to recapture Henderson Field in 1942.

W53 : Scott 326 : Marshall Islands 1992 : The Battle of El Alamein, 1942.
Designer : Brian Sanders.
Selvedge commentary :
"It may almost be said, before Alamein we never had a victory. After Alamein we never had a defeat" - Winston S Churchill.

The images shows the two senior antagonists :
Field Marshal Bernard Law Montgomery, nicknamed "Monty", a senior British Army officer who served in the First World War, the Irish War of Independence and the Second World War.

Johannes Erwin Eugen Rommel popularly known as The Desert Fox, a German Generalfeldmarschall during World War II. He served in the Wehrmacht (armed forces) of Nazi Germany, as well as in the Reichswehr of the Weimar Republic, and the army of Imperial Germany.

There were two Battles of El Alamein in World War II, both fought in 1942. The battles occurred during the North African campaign in Egypt, in and around an area named after a railway stop called El Alamein.
- First Battle of El Alamein: 1–27 July 1942
- Second Battle of El Alamein: 23 October – 4 November 1942

In addition, the Battle of Alam el Halfa (30 August – 5 September 1942) was fought during the same period and in the same location.

The Allied victory was the beginning of the end of the Western Desert Campaign, eliminating the Axis threat to Egypt, the Suez Canal and the Middle Eastern and Persian oil fields. The battle revived the morale of the Allies, being the first big success against the Axis since Operation Crusader in late 1941. The end of the battle coincided with the Allied invasion of French North Africa in Operation Torch on 8 November, which opened a second front in North Africa.

W 54 : Scott 327-328 : Marshall Islands 1992 : The Battle of the Barents Sea, 1942.

Designer : Brian Sanders.

Selvedge commentary :

" . . . *that an enemy force . . . with all the advantages of surprise and concentration, should be held off . . . without any loss to the convoy is most creditable and satisfactory*" – Admiral Sir John Tovey.

"*Break off engagement and retire westward*"- Vice Admiral Oskar Kummetz.

The two vessels named by the designer are (1) *HMS Sheffield* and (2) *Admiral Hipper*.

The Battle of the Barents Sea was a World War II naval engagement on 31 December 1942 between warships of the German Navy (*Kriegsmarine*) and British ships escorting Convoy JW 51B to Kola Inlet in the USSR. The action took place in the Barents Sea north of North Cape, Norway. The German raiders' failure to inflict significant losses on the convoy infuriated Hitler, who ordered that German naval strategy would henceforth concentrate on the U-boat fleet rather than surface ships.

Oskar Kummetz (1891 – 1980) was an admiral with the Kriegsmarine. On 1 March 1944, Kummetz became the Commander-in-Chief of Naval High Command Baltic Sea in Kiel. On 16 September 1944 he was promoted to *Generaladmiral*. In the final months of the war, Kummetz was responsible for Operation Hannibal, the evacuation of German refugees and military personnel from Courland, East Prussia, West Prussia and Pomerania through the Baltic Sea.

W 55 : Scott 329 : Marshall Islands 1993 : The Casablanca Conference, 1943.
Designer : Howard Koslow.
Selvedge commentary :
". . . we must bend ourselves to the task of weighting our blows more heavily"-
Winston S Churchill.

The Casablanca Conference was held in Casablanca, French Morocco, from
January 14 to 24, 1943, to plan the Allied European strategy for the next phase
of World War II. The main discussions were between US President Franklin
Roosevelt (with his military staff) and British Prime Minister Winston
Churchill (with his staff). Stalin could not attend. Key decisions included a
commitment to demand Axis powers' unconditional surrender; plans for an
invasion of Sicily and Italy before the main invasion of France; an intensified
strategic bombing campaign against Germany; and approval of a US Navy plan
to advance on Japan through the central Pacific and the Philippines. The last
item authorized the island-hopping campaign in the Pacific, which shortened the
war. Of all the decisions made, the most important was the Allied invasion of
Sicily, which Churchill pushed for in part to divert American attention from
opening a second front in France in 1943, a move that he feared would result in
very high Allied casualties and not be possible until 1944.
Behind the scenes, the United States and the United Kingdom were divided in
the commitment to see the war through to Germany's capitulation and
"unconditional surrender". But Churchill had agreed in advance about
"unconditional surrender"; he had cabled the War Cabinet four days earlier and
they had not objected. US General George Marshall also said that he had been
consulted; he had stated on 7 January that Allied morale would be "strengthened
by the uncompromising demand, and Stalin's suspicions allayed".

The British felt that arriving at some accommodation with Germany would allow
the German Army to help fight off a Soviet takeover of Eastern Europe. To
Churchill and the other Allied leaders, the real obstacle to realising that mutual
strategy with Germany was the leadership of Adolf Hitler. Allen Dulles, the chief

of OSS intelligence in Bern, Switzerland, maintained that the Casablanca Declaration was "merely a piece of paper to be scrapped without further ado if Germany would sue for peace. Hitler had to go." There is evidence that German resistance forces, highly placed anti-Nazi government officials, were working with British intelligence, MI6, to eliminate Hitler and negotiate a peace with the Allies. One such individual was Admiral Wilhelm Canaris, head of German intelligence, the Abwehr. His persistent overtures for support from the United States were ignored by Roosevelt.

W56 : Scott 330 : Marshall Islands 1993 : The Liberation of Kharkov, 1943.
Designer : Brian Sanders.
Selvedge commentary :
"The mass expulsion of the enemy from the Soviet Union has begun" – Joseph Stalin.

The Battle of Kharkov was any one of four World War II battles in and near the Soviet city of Kharkov in modern Ukraine. In usage the term is sometimes indistinct, perhaps meaning the collection of all fighting at Kharkov including and in between the four named battles, or perhaps meaning just one of the battles without specifying which. For example, soldiers have received awards "for their action in the Battle of Kharkov".

The four named battles are:

- First Battle of Kharkov, an October 1941 battle in which German troops captured the city.
- Second Battle of Kharkov, a May 1942 battle in which Soviet forces attempted to retake the city.
- Third Battle of Kharkov, a February 1943 battle in which Soviet forces were driven out again, and the Germans forces retook the city. This is the event commemorated with the image above.

- Belgorod–Kharkov offensive operation, an August 1943 battle in which Soviet forces retook the city. In German nomenclature, the operation is usually referred to as the Fourth Battle of Kharkov.

Joseph Vissarionovich Stalin (1878 – 1953) was the Soviet politician, revolutionary, and political theorist who led the Soviet Union from 1924 until his death in 1953.

W57 : Scott 331-334 : Marshall Islands 1993 : The Battle of Bismark Sea 1943.
Designer : Brian Sanders.
Selvedge commentary :
"Please extend to all ranks my gratitude and felicitations on the magnificent victory which has been achieved.
It cannot fail to go down in history as one of the most complete and annihilating combats of all time" – General Douglas MacArthur.

The recognised weapons are (1) Japanese Mitsubishi A6M 'Zeros' and destroyer Asagumo, (2) USAF 'Lightnings' and RAAF 'Beaufighters', (3) the Japanese destroyer Yukikasi, (4) USAF A20 'Havok' and B25 'Mitchell'.

The Battle of the Bismarck Sea took place in the South West Pacific Area during World War II when aircraft of the U.S. Fifth Air Force and the Royal Australian Air Force attacked a Japanese convoy carrying troops to Lae, New Guinea. Most of the Japanese task force was destroyed, and Japanese troop losses were heavy. The Japanese convoy was a result of a Japanese Imperial General Headquarters decision in December 1942 to reinforce their position in the South West Pacific. A plan was devised to move some 6,900 troops from Rabaul directly to Lae. The plan was understood to be risky, because Allied air power in the area was strong, but it was decided to proceed because otherwise the troops would have to be landed a considerable distance away and march through inhospitable swamp, mountain and jungle terrain without roads before reaching their destination. On 28 February 1943, the convoy – comprising eight destroyers and eight troop transports with an escort of approximately 100 fighter aircraft – set out from Simpson Harbour in Rabaul. Naval codebreakers in Melbourne (FRUMEL) and Washington, D.C., had decrypted and translated messages indicating the convoy's intended destination and date of arrival. The Allied Air Forces had developed new techniques, such as skip bombing, that they hoped would improve the chances of successful air attack on ships. They detected and shadowed the convoy, which came under sustained air attack on 2–3 March 1943. Follow-up attacks by PT boats and aircraft were made on 4 March on lifeboats and rafts. All eight transports and four of the escorting destroyers were sunk. Of 6,900 troops who were badly needed in New Guinea, only about 1,200 made it to Lae. Another 2,700 were rescued by destroyers and submarines and returned to Rabaul. The Japanese made no further attempts to reinforce Lae by ship, greatly hindering their ultimately unsuccessful efforts to stop the Allied offensives in New Guinea.

W58 : Scott 335 : Marshall Islands 1993 : The interception of Admiral
Yamamoto, 1943.

Designer : David K Stone.
Selvedge commentary :
"One could almost hear the rising crescendo of sound from . . . the bottom of Pearl Harbour" -
General Douglas MacArthur.

Operation Vengeance was the American military operation to kill
Admiral Isoroku Yamamoto of the Imperial Japanese Navy on 18 April 1943
during the Solomon Islands campaign in the Pacific Theatre. Yamamoto,
commander of the Combined Fleet of the Imperial Japanese Navy, was killed
near Bougainville Island when his G4M1 transport aircraft was shot down
by United States Army Air Forces fighter aircraft operating from Kukum
Field on Guadalcanal.

The mission of the U.S. aircraft was specifically to kill Yamamoto, made possible
because of United States Navy intelligence decoding transmissions about
Yamamoto's travel itinerary through the Solomon Islands area. The death of
Yamamoto reportedly damaged the morale of Japanese naval personnel, raised
the morale of the Allied forces, and was intended as revenge by U.S. leaders, who
blamed Yamamoto for the attack on Pearl Harbor that initiated the war
between Imperial Japan and the United States.
The U.S. pilots claimed to have shot down three twin-engine bombers and two
fighters during the mission, but Japanese records show only two bombers were
shot down. There is a controversy over which pilot shot down Yamamoto's
plane, but most modern historians credit Rex T. Barber.

On April 18, Lieutenant Barber figured prominently in the Yamamoto
interception, also known as Operation Vengeance. Intelligence sources had
learned that Yamamoto would be flying in a "Betty" bomber on an inspection
tour of Japanese bases in the northern Solomon Islands. Historian Donald P.
Bourgeois credits Barber with the sole kill of Yamamoto's aircraft. In 1991,

Barber and Captain Thomas George Lanphier Jr. were officially credited with half a kill each in Yamamoto's bomber after the Air Force reviewed the incident. Barber also shared a second Betty destroyed on the same mission.

W59 : Scott 336-337 : Marshall Islands 1993 : Battle of Kursk 1943.
Designer : Brian Sanders.
Selvedge commentary : *"The next two or three days will be terrible. 'Either we hold out or the Germans will take Kursk"*- Lieutenant General Nikita S Khrushchev.
"Around us everything was in motion . . .'We both felt and heard the hurricane of fire".
Marshall Georgi Zhukov.

The recognised weapons are (1) German 'Tiger 1' and (2) a 'Soviet T-34'.

The Battle of Kursk was a major World War II Eastern Front battle between the forces of Nazi Germany and the Soviet Union near Kursk in southwestern Russia during the summer of 1943, resulting in a Soviet victory. The Battle of Kursk is the single largest battle in the history of warfare. It ranks only behind the Battle of Stalingrad several months earlier as the most often-cited turning point in the European theatre of the war. It was one of the costliest battles of the Second World War, the single deadliest armoured battle in history, and the opening day of the battle, 5 July, was the single costliest day in the history of aerial warfare. The battle was also marked by fierce house-to-house fighting and hand-to-hand combat.

The battle began with the launch of the German offensive Operation Citadel, which had the objective of pinching off the Kursk salient with attacks on the base of the salient from north and south simultaneously. After the German offensive stalled on the northern side of the salient, on 12 July, the Soviets commenced their Kursk Strategic Offensive Operation with the launch of Operation Kutuzov against the rear of the German forces on the same side. On the southern side, the Soviets also launched powerful counterattacks the same day, one of which led to a large armoured clash, the Battle of Prokhorovka. On 3 August, the Soviets began the second phase of the Kursk Strategic Offensive Operation with the launch of the Belgorod–Kharkov offensive operation against the German forces on the southern side of the salient.

The Germans hoped to weaken the Soviet offensive potential for the summer of 1943, by cutting off and enveloping the forces that they anticipated would be in the Kursk salient. Hitler believed that a victory here would reassert German strength and improve his prestige with his allies, who he thought were considering withdrawing from the war. It was also hoped that large numbers of Soviet prisoners would be captured to be used as slave labour in the German armaments industry. The Soviet government had foreknowledge of the German plans from the Lucy spy ring. Aware months in advance that the attack would fall on the neck of the Kursk salient, the Soviets built a defence in depth designed to wear down the German armoured spearhead. The Germans delayed the offensive while they tried to build up their forces and waited for new weapons, giving the Red Army time to construct a series of deep defensive belts and establish a large reserve force for counter-offensives, with one German officer describing Kursk as "another Verdun".

The battle was the final strategic offensive that the Germans were able to launch on the Eastern Front. Because the Allied invasion of Sicily began during the battle, Adolf Hitler was forced to divert troops training in France to meet the Allied threat in the Mediterranean, rather than using them as a strategic reserve for the Eastern Front. As a result, Hitler cancelled the offensive at Kursk after only a week, in part to divert forces to Italy. Germany's heavy losses of men and tanks ensured that the victorious Soviet Red Army held a strategic initiative for the rest of the war. The Battle of Kursk was the first time in the Second World War that a German strategic offensive was halted before it could break through enemy defences and penetrate to its strategic depths. Though the Red Army had succeeded in winter offensives previously, their counter-offensives after the

German attack at Kursk were their first successful summer offensives of the war. The battle has been called the "last gasp of Nazi aggression".

W60 : Scott 467-470 : Marshall Islands 1993 : The Allied invasion of Sicily, 1943. █

Designer : David K Stone.

Selvedge commentary :

"This is a horse race in which the prestige of the US Army is at stake, we must take Messina before the British" - Lieutenant General George S Patton.

"My battle situation is very good ... Suggest my Army operates offensively northwards to cut the Island in two" - General Bernard L. Montgomery.

The two portraits are named : General George Patton and General Bernard Montgomery.

The Battle paintings are : "Americans landing at Licata", "British Landing south of Syracuse".

The Allied invasion of Sicily, codenamed Operation *Husky*, was a major battle of World War II in which Allies captured Sicily from the Axis Powers (Italy and Nazi Germany).

The large sea and air operation was followed by six weeks of land fighting and started the Italian Campaign. It was followed by the Allied invasion of Italy.

Husky began on the night of 9–10 July 1943 and ended on 17 August. It achieved the goals of the Allies. The Allies removed Axis air, land, and naval forces from the island.

As well, the Mediterranean's sea lanes were opened to the Allies, and the Italian dictator, Benito Mussolini, was briefly removed from power.

---

[3] You will note the Scott numbering jumps here, reason unknown.

The plan for Operation *Husky* called for the sea attack of the island by two armies. One would land on the southeastern coast. Another would land on the central southern coast. The attack would be helped by naval gunfire and bombing, and its commander was US General Dwight D. Eisenhower.

The Allied land forces were from the American, British, and Canadian Armies and were put into two groups. The Eastern Task Force was led by General Bernard Montgomery. The Western Task Force was commanded by Lieutenant General George S. Patton. In addition to the sea landings, airborne troops were to be flown in to capture bridges and high ground.

The island was defended by the two corps of Italian 6th Army under General Alfredo Guzzoni. In early July, the total Axis force in Sicily was about 200,000 Italian and 32,000 German troops, and 30,000 *Luftwaffe* ground staff. By late July, there were 70,000 German troops.

The attack plan was made on 17 May. At the Casablanca Conference in January 1943, American and British political and military leaders met to discuss future plans. The British wanted an invasion of Sicily.

George Smith Patton, Junior (1885 – 1945) was the general in the United States Army who commanded the Seventh Army in the Mediterranean Theatre of World War II, then the Third Army in France and Germany after the Allied invasion of Normandy in June 1944.

Field Marshal Bernard Law Montgomery, 1st Viscount Montgomery of Alamein (1887 – 1976), nicknamed "Monty", was a senior British Army officer who served in the First World War, the Irish War of Independence and the Second World War.
With the Allied invasion of Sicily (Operation Husky). Montgomery considered the initial plans for the Allied invasion, which had been agreed in principle by General Dwight D. Eisenhower, the Supreme Allied Commander Allied Forces Headquarters, and General Alexander, the 15th Army Group commander, to be unworkable because of the dispersion of effort. He managed to have the plans recast to concentrate the Allied forces, having Lieutenant General George Patton's US Seventh Army land in the Gulf of Gela (on the Eighth Army's left flank, which landed around Syracuse in the south-east of Sicily) rather than near Palermo in the west and north of Sicily. Inter-Allied tensions grew as the American commanders, Patton

and Omar Bradley (then commanding US II Corps under Patton), took umbrage at what they saw as Montgomery's attitudes and boastfulness. However, while they were considered three of the greatest soldiers of their time, due to their competitiveness they were renowned for "squabbling like three schoolgirls" thanks to their "bitchiness", "whining to their superiors" and "showing off".

W61 : Scott 471 : Marshall Islands 1993 : Bomber raids on Schweinfurt, 1943.
Designer : Brian Sanders.
Selvedge commentary :
"It was a bold strategic concept, one of the most significant and remarkable air battles of the Second World War . . ." – Lieutenant General Ira C Eaker.

The Schweinfurt-Regensburg raids were a series of attacks by B-17 Flying Fortress and B-24 Liberator bombers of the United States Air Force in August and October 1943. The Eighth Air Force attack against the ball bearing factories at Schweinfurt on October 14, 1943, became known as "Black Thursday". The Schweinfurt–Regensburg mission was a strategic bombing mission during World War II carried out by Boeing B-17 Flying Fortress heavy bombers of the US Army Air Forces on August 17, 1943. The Eighth Air Force made its second attempt to destroy the ball-bearing plants of Schweinfurt on October 14, 1943, which largely accomplished its goals but proved so costly that further operations ceased until a new model of attack could be formulated.

Ira Clarence Eaker (1896 - 1987) was a general of the United States Army Air Forces. Eaker, as second-in-command of the prospective Eighth Air Force, was sent to England to form and organize its bomber command. While he struggled to build up airpower in England, the organization of the Army Air Forces evolved, and he was named commander of the Eighth Air Force on December 1, 1942.
Although his background was in single-engine fighter aircraft, Eaker became the architect of a strategic bombing force that ultimately numbered forty groups of 60 heavy bombers each, supported by a subordinate fighter command of 1,500 aircraft, most of which was in place by the time he relinquished command at the start of 1944. Eaker then took overall command of four Allied air forces based in the Mediterranean Theatre of Operations, and by the end of World War II had been named Deputy Commander of the U.S. Army Air Forces.

W62 : Scott 472 : Marshall Islands 1993 : Liberation of Smolensk, 1943.
Designer : Shannon Sternweis.
Selvedge commentary :
"When in danger . . . (the Germans) flee, abandoning their equipment and their wounded on the field"- Joseph Stalin, November 7, 1943.

During World War II, Smolensk saw wide-scale fighting during the first Battle of Smolensk when the city was captured by the Germans on 16 July 1941. The first Soviet counteroffensive against the German army was launched in August but failed. However, the limited Soviet victories outside the city halted the German advance for a crucial two months, granting time to Moscow's defenders to prepare in earnest. Over 93% of the city was destroyed during the fighting; the ancient icon of Our Lady of Smolensk was lost. Nevertheless, it escaped total destruction. In late 1943, Hermann Göring had ordered Gotthard Heinrici to destroy Smolensk in accordance with the Nazi "scorched earth" policy. He refused and was punished for it. The city was finally liberated on 25 September 1943, during the second Battle of Smolensk. The rare title of Russian Hero City was bestowed on Smolensk after the war.

Despite an impressive German defence, the Red Army was able to stage several breakthroughs, liberating several major cities, including Smolensk and Roslavl. As a result of this operation, the Red Army was able to start planning for the liberation of Belarus. However, the overall advance was quite modest and slow in the face of heavy German resistance, and the operation was therefore accomplished in three stages.

Although playing a major military role, the Smolensk operation was also important for its effect on the Battle of the Dnieper. It has been estimated that as many as 55 German divisions were committed to counter the Smolensk operation – divisions which would have been critical to prevent Soviet troops from crossing the Dnieper in the south. During the operation, the Red Army also definitively drove back German forces from the Smolensk land bridge, historically the most important approach for a western attack on Moscow.

W63 : Scott 473 : Marshall Islands 1993 : Landing at Bougainville. 1943.
Designer : Shannon Sternweis.
Selvedge commentary :
"This is the hottest potato they ever handed me"- Admiral William (Bull) Halsey,
October 18, 1943.

The Bougainville campaign was a series of land and naval battles of the Pacific
campaign between Allied forces and the Empire of Japan, named after the island
of Bougainville. The campaign took place in the Northern Solomons in two
phases. The first phase, in which American troops landed and held the perimeter
around the beachhead at Torokina, lasted from November 1943 through
November 1944. The second phase, in which primarily Australian troops went
on the offensive, mopping up pockets of starving, isolated but still-determined
Japanese, lasted from November 1944 until August 1945, when the last Japanese
soldiers on the island surrendered. Operations during the final phase of the
campaign saw the Australian forces advance north towards the Bonis
Peninsula and south towards the main Japanese stronghold around Buin,
although the war ended before these two enclaves were completely destroyed.

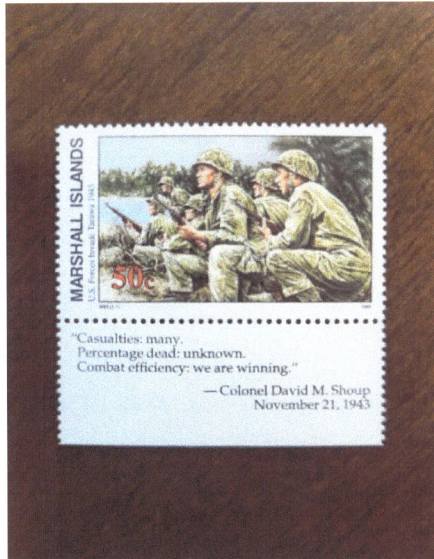

W 64 : Scott 474 : Marshall Islands 1993 : U S Forces invade Tarawa, 1943.
Selvedge commentary :
"Casualties many.
Percentage dead : unknown.
Combat efficiency : we are winning" – Colonel David M Shoup, November 21, 1943.

To set up forward air bases capable of supporting operations across the Central Pacific, to the Philippines, and into Japan, the U.S. planned to take the Mariana Islands. The Marianas were heavily defended. Naval doctrine of the time held that in order for attacks to succeed, land-based aircraft would be required to weaken the defences and protect the invasion forces. The nearest islands capable of supporting such an effort were the Marshall Islands. Taking the Marshalls would provide the base needed to launch an offensive on the Marianas, but the Marshalls were cut off from direct communications with Hawaii by a Japanese garrison and air base on the small island of Betio, on the western side of the atoll of Tarawa in the Gilbert Islands. Thus, to eventually launch an invasion of the Marianas, the battle had to start far to the east at Tarawa.

The Battle of Tarawa was fought on 20–23 November 1943 between the United States and Japan at the Tarawa Atoll in the Gilbert Islands, and was part of Operation Galvanic, the U.S. invasion of the Gilberts. Nearly 6,400 Japanese, Koreans, and Americans died in the fight, mostly on and around the small island of Betio, in the extreme southwest of Tarawa Atoll. At the time, Betio was only 118 hectares (290 acres).

The Battle of Tarawa was the first American offensive in the critical Central Pacific region. It was also the first time in the Pacific War that the United States faced serious Japanese opposition to an amphibious landing. Previous landings had met little to no initial resistance, but on Tarawa the 4,500 Japanese defenders were well supplied and well prepared, and they fought almost to the last man, exacting a heavy toll on the United States Marine Corps.

David Monroe Shoup (1904 – 1983) who became a general of the United States Marine Corps was awarded the Medal of Honour served as the 22nd Commandant of the Marine Corps. He served in the Marianas campaign, and later became a high-level military logistics officer.

W65 : Scott 475 : Marshall Islands 1993 : The Tehran Allies Conference, 1943.
Designer : Shannon Sternweis.
Selvedge commentary :
"Our nations shall work together in the war and in the peace that will follow"- The three
power declaration", December 1, 1943.

The Tehrān Conference, (November 28–December 1, 1943), meeting
between U.S. President Franklin D. Roosevelt, British Prime Minister Winston
Churchill, and Soviet Premier Joseph Stalin in Tehrān. The chief discussion
centred on the opening of a "second front" in western Europe. Stalin agreed to
an eastern offensive to coincide with the forthcoming Western Front, and he
pressed the western leaders to proceed with formal preparations for their long-
promised invasion of German-occupied France.

Though military questions were dominant, the Tehrān Conference saw more
discussion of political issues than had occurred in any previous meeting
between the Allied governmental heads. Not only did Stalin reiterate that
the Soviet Union should retain the frontiers provided by the German-Soviet
Nonaggression Pact of 1939 and by the Russo-Finnish Treaty of 1940, but he
also stated that it would want the Baltic coast of East Prussia. Though the
settlement for Germany was discussed at length, all three Allied leaders appeared
uncertain; their views were imprecise on the topic of a postwar international
organization; and, on the Polish question, the western Allies and the Soviet
Union found themselves in sharp dissension, Stalin expressing his continued
distaste for the Polish government-in-exile in London. On Iran,
which Allied forces were partly occupying, they were able to agree on a
declaration (published on December 1, 1943) guaranteeing the postwar
independence and territorial integrity of that state and promising postwar
economic assistance.

W66 : Scott 476 : Marshall Islands 1993 : The Battle of North Cape, 1943.
Designer : Brian Sanders.
Selvedge commentary :
"Finish her off with torpedoes"- Admiral Sir Bruce Austin Fraser, December 26, 1943.
"Unbelievable scenes took place as the heavy shells of the British battleships poured into the doomed Scharnhorst"- Scharnhorst survivor, December 30, 1943.

The two battleships illustrated are the *Scharnhorst* and *HMS Duke of York*.

Since August 1941, the western Allies had run convoys of ships from the United Kingdom and Iceland to the northern ports of the Soviet Union to provide essential supplies for their war effort on the Eastern Front. These endured much hardship, frequently attacked by German naval and air forces stationed in occupied Norway. A key concern were German *Kriegsmarine* battleships such as *Tirpitz* and *Scharnhorst*. Even the threat of these ships' presence was enough to cause disastrous consequences for the convoys, such as Convoy PQ 17 that was scattered and mostly sunk by German forces after false reports of the *Tirpitz* sailing to intercept them. To ward off the threat of Germany's capital ships in the Arctic and to escort convoys with a high level of success, the Royal Navy had to outlay great assets.

The Battle of the North Cape, Barents Sea occurred on 26 December 1943, as part of the Arctic campaign. The German battleship *Scharnhorst*, on an operation to attack Arctic convoys of war materiel from the western Allies to the Soviet Union, was brought to battle and sunk by the Royal Navy's battleship HMS *Duke of York* with cruisers and destroyers, including an onslaught from the destroyer HNoMS *Stord* of the exiled Royal Norwegian Navy, off the North Cape, Norway.

The sinking of the *Scharnhorst* was a major victory for the Allied war effort in the Arctic theatre and further altered the strategic balance at sea in their favour. The Battle of the North Cape took place only a few months after the successful Operation Source, which had severely damaged the German battleship *Tirpitz* with midget submarines as she lay at anchor in Norway. With *Scharnhorst* destroyed and Germany's other battleships out of service, the Allies were now for the first time in the war free from the threat of German battleships raiding their convoys in the Arctic and Atlantic. This would allow the Allies to reallocate their naval resources that had been previously tied up to counter the threat of the German 'fleet in being'. This would prove to be the final battle of battleships in European waters and was one of few major surface actions in the Second World War without air support.

At the start of World War II Admiral Sir Bruce Austin Fraser, was controller of the navy and was in large part responsible for directing its expansion during the 1939–41 period. Fraser then became commander in chief of the Home Fleet and was chiefly concerned with the protection of convoys to the U.S.S.R. On Dec. 26, 1943, aboard his flagship "Duke of York," he fought and sank the German battleship "Scharnhorst" off Norway's North Cape in an engagement conducted largely at night with the aid of radar. As admiral in 1944 he was appointed commander in chief of the British Pacific fleet and on Sept. 2, 1945, signed the Japanese surrender papers for Great Britain in Tokyo Bay.

W67 : Scott 478 : Marshall Islands 1994 : The appointment of General Dwight D. Eisenhower as Commander of Supreme Headquarters, Allied Expeditionary Force, 1944.

Designer : Howard Koslow.

Selvedge commentary :

"Mr. President, I realise that such an appointment involved difficult decisions. I hope you will not be disappointed" - General Dwight D Eisenhower, December 7, 1943.

During World War II, General Eisenhower was Supreme Commander of the Allied Expeditionary Force in Europe and achieved the five-star rank as General of the Army. Eisenhower planned and supervised two of the most consequential military campaigns of World War II: Operation Torch in the North Africa campaign in 1942–1943 and the invasion of Normandy in 1944.

After the war ended in Europe, he served as military governor of the American-occupied zone of Germany (1945), Army Chief of Staff (1945–1948), president of Columbia University (1948–1953), the first supreme commander of NATO (1951–1952), and President of the USA.

W 68 : Scott 479 : Marshall Islands 1994 : The invasion of Anzio, 1944. Designer : David K Stone.

Selvedge commentary :

"*We achieved what is certainly one of the most complete surprises in history*"- Major General John P Lucas, January 22, 1944.

The Battle of Anzio was a battle of the Italian Campaign of World War II that commenced January 22, 1944. The battle began with the Allied amphibious landing known as Operation Shingle, and ended on June 4, 1944, with the liberation of Rome. The operation was opposed by German and by Italian *Repubblica Sociale Italiana* (RSI) forces in the area of Anzio and Nettuno.

The operation was initially commanded by Major General John P. Lucas, of the U.S. Army, commanding U.S. VI Corps with the intent to outflank German forces at the Winter Line and enable an attack on Rome. Major General John P. Lucas After the initial success of the landings at Anzio on January 22, and with little German resistance in the area, Lucas had the opportunity to break out of the beachhead and cut off the supply lines of the German 10th Army by crossing Highways 6 and 7, leaving the way open to Rome. He failed to seize the opportunity, deciding instead to wait until all of his ground troops had landed and the beachhead had been fully secured.

The success of an amphibious landing at that location, in a basin consisting substantially of reclaimed marshland and surrounded by mountains, depended on the element of surprise and the swiftness with which the invaders could build up strength and move inland relative to the reaction time and strength of the defenders. Any delay could result in the occupation of the mountains by the defenders and the consequent entrapment of the invaders. Lieutenant General Mark W. Clark, commander of the U.S. Fifth Army, understood that risk, but he did not pass on his appreciation of the situation to his subordinate Lucas,[citation needed] who preferred to take time to entrench against an expected counterattack. The initial landing achieved complete surprise with no opposition and a jeep patrol even made it as far as the outskirts of Rome. However, Lucas, who had little confidence in the operation as planned, failed to capitalize on the element of surprise and delayed his advance until he judged his position was sufficiently consolidated and he had sufficient strength.

While Lucas consolidated, Field Marshal Albert Kesselring, the German commander in the Italian theatre, moved every unit he could spare into a defensive ring around the beachhead. His artillery units had a clear view of every Allied position. The Germans also stopped the drainage pumps and flooded the reclaimed marsh with salt water, planning to entrap the Allies and destroy them by epidemic. For weeks a rain of shells fell on the beach, the marsh, the harbour, and on anything else observable from the hills, with little distinction between forward and rear positions.

After a month of heavy but inconclusive fighting, Lucas was relieved and sent home. His replacement was Major General Lucian Truscott, who had commanded the U.S. 3rd Infantry Division. The Allies broke out in May. But, instead of striking inland to cut lines of communication of the German Tenth Army's units fighting at Monte Cassino, Truscott, on Clark's orders, reluctantly turned his forces north-west towards Rome, which was captured on June 4, 1944. As a result, the forces of the German Tenth Army fighting at Cassino were able to withdraw and rejoin the rest of Kesselring's forces north of Rome, regroup, and make a fighting withdrawal to his next major prepared defensive position on the Gothic Line.

The battle was costly, with 24,000 U.S. and 10,000 British casualties.

Albert Kesselring (1885 – 1960) was a German military officer and convicted war criminal who served in the *Luftwaffe*. In a career which spanned both world wars, Kesselring reached the rank of the *Generalfeldmarschall* (Field marshal) and became one of Nazi Germany's most highly decorated commanders.

General Lucian King Truscott Jr. (1895 – 1965) was a highly decorated senior United States Army officer, who saw distinguished active service. Between 1943–1945, he successively commanded the 3rd Infantry Division, VI Corps, Fifteenth Army and Fifth Army, serving mainly in the Mediterranean Theatre of Operations (MTO) during his wartime service.

W69 : Scott 480 : Marshall Islands 1944 : The Siege of Leningrad lifted, 1944.
Designer : Brian Sanders.
Selvedge commentary :
"Never in the history of the world, has there been an example of tragedy to equal that of starving Leningrad"- Official Leningrad History '*The 900 days*'.

The siege of Leningrad was a prolonged military siege undertaken by the Axis powers against the city of *Leningrad* (present-day Saint Petersburg) on the Eastern Front . Germany's Army Group North advanced from the south, while the German-allied Finnish army invaded from the north and completed the ring around the city.

The siege began on 8 September 1941, when the Wehrmacht severed the last road to the city. Although Soviet forces managed to open a narrow land corridor to the city on 18 January 1943, the Red Army did not lift the siege until 27 January 1944, 872 days (125 weeks) after it began. The siege became one of the longest and most destructive sieges in history, and it was possibly the costliest siege in history due to the number of casualties which were suffered throughout its duration. An estimated 1.5 million people died as a result of the siege. At the time, it was not classified as a war crime, however, in the 21st century, some historians have classified it as a genocide, due to the intentional destruction of the city and the systematic starvation of its civilian population.

W70 : Scott 481 : Marshall Islands 1944 : US liberates the Marshall Islands, 1944.
Designer Howard Koslow.
Selvedge commentary :
"To all hands concerned with the Marshalls . . . Well and smartly done. Carry on" –
Admiral Ernest J King, February 1944.

The aircraft is a Douglas SBD Dauntless, an American naval scout
plane and dive bomber that was manufactured by Douglas Aircraft from 1940
through 1944. The SBD ("Scout Bomber Douglas") was the United States
Navy's main carrier-based scout/dive bomber from mid-1940 through mid-1944.
The SBD was also flown by the United States Marine Corps, both from land air
bases and aircraft carriers. The SBD is best remembered as the bomber that
delivered the fatal blows to the Japanese carriers at the Battle of Midway in June
1942. The type earned its nickname "Slow But Deadly" (from its SBD initials)
during this period.

The Battle of Kwajalein was fought as part of the Pacific campaign of World
War II. It took place 31 January – 3 February 1944, on Kwajalein Atoll in
the Marshall Islands. Employing the hard-learned lessons of the Battle of
Tarawa, the United States launched a successful twin assault on the main islands
of Kwajalein in the south and Roi-Namur in the north. The Japanese defenders
put up stiff resistance, although outnumbered and under-prepared. The
determined defence of Roi-Namur left only 51 survivors of an original garrison
of 3,500.

For the US, the battle represented both the next step in its island-hopping march
to Japan and a significant morale victory because it was the first time the
Americans had penetrated the "outer ring" of the Japanese Pacific sphere. For
the Japanese, the battle represented the failure of the beach-line defence.
Japanese defences became prepared in depth, and the Battles of Peleliu, Guam,
and the Marianas proved far more costly to the US.

W71 : Scott 482 : Marshall Islands 1994 : Japanese defeated at Truk, 1944.
Designer : Shannon Stirnweis.
Selvedge commentary :
"The Pacific Fleet has returned at Truk the visit made by the Japanese at Pearl Harbour"-
Admiral Chester Nimitz, February 17, 1944.

I am not totally positive in suggesting the two background figures in the Stirnweis painting are Admiral Raymond Spruance and Vice Admiral Mark Mitscher.

With their position in the Solomons disintegrating, the Japanese modified the Z Plan by eliminating the Gilbert and Marshall Islands and the Bismarck's as vital areas to be defended. They then based their possible actions on the defence of an inner perimeter, which included the Marianas, Palau, Western New Guinea, and the Dutch East Indies. Meanwhile, in the Central Pacific a major American offensive was initiated, beginning in November 1943 with landings in the Gilbert Islands. The Japanese were forced to watch helplessly as their garrisons in the Gilberts and then the Marshalls were crushed. The Japanese strategy of holding overextended island garrisons was fully exposed.

Because aircraft stationed at Truk could potentially interfere with the upcoming invasion of Eniwetok, and because Truk had recently served as a ferry point for the resupply of aircraft to Rabaul, Admiral Raymond Spruance ordered Vice Admiral Marc Mitscher's Fast Carrier Task Force, designated TF 58, to carry out air raids against Truk.

In February 1944, the US Navy's TF58 attacked during Operation Hailstone. Although the Combined Fleet had moved its major vessels out in time to avoid being caught at anchor in the atoll, two days of air attacks resulted in significant losses to Japanese aircraft and merchant shipping. The power of the American attack on Truk far surpassed that of the Japanese attack against Pearl Harbor. The

IJN was forced to abandon Truk and was now unable to stop the Americans on any front. Consequently, the Japanese retained their remaining strength in preparation for what they hoped would be a decisive battle.

W72 : Scott 483 : Marshall Islands 1994 : Big week - US bombs Germany, 1944.
Designer : David K Stone.
Selvedge commentary :
"The operation set a record for size . . . 'Those five days changed the history of the air war"-
General Henry H Arnold.

Operation Argument, after the war was dubbed *Big Week*. It was a sequence of raids by the United States Army Air Forces and RAF Bomber Command from 20 to 25 February 1944, as part of the Combined Bomber Offensive against Nazi Germany. The objective of Operation Argument was to destroy aircraft factories in central and southern Germany in order to defeat the *Luftwaffe* before the Normandy landings during Operation Overlord were to take place later in 1944.

The joint daylight bombing campaign was also supported by RAF Bomber Command operating against the same targets at night. Arthur "Bomber" Harris resisted contributing RAF Bomber Command so as not to dilute the British "area bombing" offensive against Berlin. It took an order from Air Chief Marshal Sir Charles Portal, Chief of the Air Staff, to force Harris to comply.

RAF Fighter Command also provided escort for USAAF bomber formations, just at the time that the Eighth Air Force had started introducing the improved long range P-51 Mustang fighter which gave the USAAF bomber forces more cover deeper into Germany, to take over the role. The offensive overlapped the German Operation Steinbock, the *Baby Blitz*, which lasted from January to May 1944.

Henry Harley Arnold (1886 –1950 General of the Army and later, General of the Air Force. Arnold was an aviation pioneer, Chief of the Air Corps (1938–1941), commanding general of the United States Army Air Forces, the only United States Air Force general to hold five-star rank, and the only officer to hold a five-star rank in two different U.S. military services.

W73 : Scott 484 : Marshall Islands 1994 : Rome falls to the Allies, 1944.
Desgner : Shannon Stirnweis.
Selvedge commentary :
"There were gay crowds in the streets, many of them waving flags, as our infantry marched through the capital"- Lieutenant General Mark Clark.

Featuring the portrait of General Mark Clark. [The General is not wearing a funny hat, he is positioned under an arch].

On 4 June 1944, the American 5th Army liberated Rome – one of the most symbolic but largely overlooked achievements of the War.

This was the first Allied unit to liberate a capital city from Fascist control on European soil. But coming just two days before D-Day, it was quickly overshadowed by the invasion of Normandy.

The fall of Rome marked the end of a bloody and protracted series of battles to break through the Germans' defensive Gustav line, principally Monte Cassino and Anzio.
Despite fierce resistance on its outskirts, German troops had been ordered to withdraw from the city itself, allowing it to be captured without any fighting.
Even though Rome was seized intact, history has not looked kindly on its liberator, the American commander of the 5th Army, Lieutenant General Mark Clark.
Controversially, Clark chose to push for Rome from the Anzio beachhead rather than cut off the retreating German forces as he had been ordered by General Sir Harold Alexander, the British officer in overall charge in Italy.

Although Rome was liberated, the Germans were not decisively defeated.

German forces fell back to a defensive line running across Italy just north of Florence, the Gothic Line. The Allies did not breach this line until September 1944.

The Allied front then stalled again until a breakthrough was made in April 1945, leading to the eventual surrender of German forces in Italy on 2 May – two days after the fall of Berlin.

The Italian campaign had tied down more than 20 German divisions while the Allies concentrated on the battle on the western front.

W74 : Scott 485-488 : Marshall Islands 1994 : D-Day / Allied landings at
Normandy, France, 1944.

Designer : Brian Sanders.

Selvedge commentary :

"... *let us all beseech the blessings of Almighty God upon this great and noble undertaking*"-
General Dwight D Eisenhower, June 6, 1944.

"... *during the early hours of this morning the first of a series of landings in force upon the
European continent has taken place*"- Winston S Churchill.

The recognised weapons shown are described as :
  (1) Horsa gliders, (2) US P51D Mustangs, British Hurricanes, (3) German
       defences, and (4) Allied amphibious landings.

I was aware that this sheet / block of four images was reissued. Following the
advice of Lisa of the Brookman Stamp Company, who suggested to me that the
reissues were because of "errors", I have acquired a copy of the re-issue. Stamp
#(2) is described differently. Now it reads *British Typhoon 1B and US P51D
Mustangs*.

Interestingly Unicover have taken the opportunity to change text associated with
three of the four images :
  (1) Horsa gliders now reads *Horsa gliders and parachute troops*,
  (2) Now reads British Typhoon 1B and US P51D Mustangs,
  (3) German defences changed to state *German defences and Pointe du Hoc*,
  (4) Same description: *Allied amphibious landings*.

Whilst considering what the errors might be to prompt the reissue I looked up whether the Hawker Hurricane fighter was used in Operation Overlord. The Britannica "chatbox" included this observation :

> "The Hawker Hurricane was a significant fighter aircraft during World War II, particularly noted for its role in the Battle of Britain and other early war engagements. However, there no specific mention of the Hurricane being used in the Battle of Normandy. The Hurricane was primarily used. in earlier stages of the war and in various theatres, including North Africa and the defence of Malta, but it was largely superseded by other aircraft by the time of the Normandy Invasion in June 1944. Therefore, it is unlikely that Hawker Hurricanes played a significant role in the Battle of Normandy of Normandy in the provided search results" (www.britannica.com).

The Normandy landings were the landing operations and associated airborne operations on 6 June 1944 of the Allied invasion of Normandy in Operation Overlord during the Second World War. It is the largest seaborne invasion in history. The operation began the liberation of France, and the rest of Western Europe, and laid the foundations of the Allied victory on the Western Front. Planning for the operation began in 1943. In the months leading up to the invasion, the Allies conducted a substantial military deception, codenamed Operation Bodyguard, to mislead the Germans as to the date and location of the main Allied landings. The weather on the day selected for D-Day was not ideal, and the operation had to be delayed 24 hours; a further postponement would have meant a delay of at least two weeks, as the planners had requirements for the phase of the moon, the tides, and time of day, that meant only a few days each month were deemed suitable. Adolf Hitler placed Field Marshal Erwin Rommel in command of German forces and developing fortifications along the Atlantic Wall in anticipation of an invasion. US President Franklin D. Roosevelt placed Major General Dwight D. Eisenhower in command of Allied forces.

W75 : Scott 489 : Marshall Islands 1994 : V-1 bombardment of England begins, 1944.

Designer : Brian Sanders.

Selvedge commentary :

"*The flying bomb is a weapon literally and essentially indiscriminate . . .* " – Winston S Churchill, July 6, 1944.

The V-1 was the first of the *Vergeltungswaffen* (V-weapons) deployed for the terror bombing of London. It was developed at Peenemünde Army Research Center in 1939 by the *Luftwaffe* at the beginning of the Second World War, and during initial development was known by the codename "Cherry Stone". Due to its limited range, the thousands of V-1 missiles launched into England were fired from launch facilities along the French (Pas-de-Calais) and Dutch coasts or by modified He 111 aircraft.

The Wehrmacht first launched the V-1s against London on 13 June 1944, one week after (and prompted by) the successful Allied landings in France. At peak, more than one hundred V-1s a day were fired at southeast England, 9,521 in total, decreasing in number as sites were overrun until October 1944, when the last V-1 site in range of Britain was overrun by Allied forces. After this, the Germans directed V-1s at the port of Antwerp and at other targets in Belgium, launching a further 2,448 V-1s. The attacks stopped only a month before the war in Europe ended, when the last launch site in the Low Countries was overrun on 29 March 1945.

As part of operations against the V-1, the British operated an arrangement of air defences, including anti-aircraft guns, barrage balloons, and fighter aircraft, to intercept the bombs before they reached their targets, while the launch sites and underground storage depots became targets for Allied attacks including strategic bombing.

In 1944 a number of tests of this weapon were apparently conducted in Tornio, Finland. On one occasion, several Finnish soldiers saw a German plane launch what they described as a bomb shaped like a small, winged aircraft. The flight and impact of another prototype was seen by Finnish frontline soldiers; they noted that its engine stopped suddenly, causing the V-1 to descend sharply, and explode on impact, leaving a crater 20–30 metres (66–98 ft) wide. These V-1s became known to Finnish soldiers as "flying torpedoes".

[I personally remember the V1. My mother and we two boys lived in a Morrison shelter in the garden in Chadwell Heath, Essex, that Dad had built before joining the Royal Air Force in 1940].

W76 : Scott 490 : Marshall Islands 1994 : US Marines land on Saipan, 1944.
Designer : Shannon Stirnweis.
Selvedge commentary :
*"We have learnt how to pulverize atolls, but now we are up against mountains and caves . . .
"*- Lieutenant General Holland M 'Howling Mad' Smith.

The Battle of Saipan was an amphibious assault launched by the United
States against the Empire of Japan during the Pacific campaign of World War
II between 15 June and 9 July 1944. The initial invasion triggered the Battle of
the Philippine Sea, which effectively destroyed Japanese carrier-based airpower,
and the battle resulted in the American capture of the island. Its occupation put
the major cities of the Japanese home islands within the range of B-29 bombers,
making them vulnerable to strategic bombing by the United States Army Air
Forces. It also precipitated the resignation of Hideki Tōjō, the prime minister of
Japan.
Saipan was the first objective in Operation Forager, the campaign to occupy
the Mariana Islands that got underway at the same time the Allies were invading
France in Operation Overlord. After a two-day naval bombardment, the
U.S. 2nd Marine Division, 4th Marine Division, and the Army's 27th Infantry
Division, commanded by Lieutenant General Holland Smith, landed on the
island and defeated the 43rd Infantry Division of the Imperial Japanese Army,
commanded by Lieutenant General Yoshitsugu Saitō. Organized resistance
ended when at least 3,000 Japanese soldiers died in a mass *gyokusai* attack, and
afterward about 1,000 civilians committed suicide.
The capture of Saipan pierced the Japanese inner defence perimeter and forced
the Japanese government to inform its citizens for the first time that the war was
not going well. The battle claimed more than 46,000 military casualties and at
least 8,000 civilian deaths. The high percentage of casualties suffered during the
battle influenced American planning for future assaults, including the
projected invasion of Japan.

Holland McTyeire (1882 – 1967), a general in the United States Marine Corps. He is sometimes called the "father" of modern U.S. amphibious warfare. In August 1942, the general took command of the Amphibious Corps, Pacific Fleet, under which he completed the amphibious indoctrination of the 2d and 3d Marine Divisions before they went overseas, and the 7th Army Division and other units involved in the Aleutians operation. The Amphibious Corps, Pacific Fleet, was later redesignated the V Amphibious Corps, and in September 1943, as commander of that unit, Smith arrived at Pearl Harbor to begin planning for the Gilbert Islands campaign.

W77 : Scott 491 : Marshall Islands 1944 : The First Battle of the Philippines Sea, 1944.

Designer : Brian Sanders.

Selvedge commentary :

"It was as easy as shooting turkeys" - 'Hellcat' pilot.

The featured aircraft in the Saunders painting is the Grumman F6F-3 Hellcat, an American carrier-based fighter aircraft Designed to replace the earlier F4F Wildcat and to counter the Japanese Mitsubishi A6M Zero, it was the United States Navy's dominant fighter in the second half of the Pacific War.

The Battle of the Philippine Sea was a major naval battle of World War II on 19–20 June 1944 that eliminated the Imperial Japanese Navy's ability to conduct large-scale carrier actions. It took place during the United States' amphibious invasion of the Mariana Islands during the Pacific War. The battle was the last of five major "carrier-versus-carrier" engagements between American and Japanese naval forces, and pitted elements of the United States Navy's Fifth Fleet against ships and aircraft of the Imperial Japanese Navy's Mobile Fleet and nearby island garrisons. This was the largest carrier-to-carrier battle in history, involving 24 aircraft carriers, deploying roughly 1,350 carrier-based aircraft.

The aerial part of the battle was nicknamed the Great Marianas Turkey Shoot by American aviators for the severely disproportional loss ratio inflicted upon Japanese aircraft by American pilots and anti-aircraft gunners. During a debriefing after the first two air battles, a pilot from USS *Lexington* remarked "Why, hell, it was just like an old-time turkey shoot down home!" The outcome is generally attributed to a wealth of highly trained American pilots with superior tactics and numerical superiority, and new anti-aircraft ship defensive technology (including the top-secret anti-aircraft proximity fuze), versus the Japanese use of replacement pilots with not enough flight hours in training and little or no combat experience. Furthermore the Japanese defensive plans were directly obtained by the Allies from the plane wreckage of the commander-in-chief of the Imperial Japanese Navy's Combined Fleet, Admiral Mineichi Koga, in March 1944.

During the battle, American submarines torpedoed and sank two of the largest Japanese fleet carriers taking part in the battle. The American carriers launched a protracted strike, sinking one light carrier and damaging other ships, but most of the American aircraft returning to their carriers ran low on fuel as night fell. Eighty American planes were lost. Although at the time the battle appeared to be a missed opportunity to destroy the Japanese fleet, the Imperial Japanese Navy had lost the bulk of its carrier air strength and would never recover. This battle, along with the Battle of Leyte Gulf four months later, marked the end of Japanese aircraft carrier operations. The few surviving carriers remained mostly in port thereafter.

Following the death of Admiral Isoroku Yamamoto on April 18, 1943, Mineichi Koga succeeded Yamamoto as Commander in Chief of the Combined Fleet. His flagship was the battleship *Musashi*. Koga attempted to revitalize Japanese naval operations by reorganization of the Combined Fleet into task forces built around aircraft carriers in imitation of the United States Navy, and organized a land-based naval air fleet to work in coordination with the carriers.

W78 : Scott 492 : Marshall Islands 1994 : US liberates Guam, 1944.
Designer : David K Stone.
Selvedge commentary :
"We are trained; we are ready; and we are going into close action"- Rear Admiral W L Ainsworth, July 21, 1944.

The Battle of Guam (21 July – 10 August 1944) was the American recapture of the Japanese-held island of Guam, a U.S. territory in the Mariana Islands captured by the Japanese from the United States in the First Battle of Guam. The battle was a critical component of Operation Forager. The recapture of Guam and the broader Mariana and Palau Islands campaign resulted in the destruction of much of Japan's naval air power and allowed the United States to establish large airbases from which it could bomb the Japanese home islands with its new strategic bomber, the Boeing B-29 Superfortress.

Guam was turned into a base for Allied operations after the battle. Five large airfields were built by the Navy Seabees and African American Aviation Engineering Battalions. Army Air Forces B-29 bombers flew from Northwest Field and North Field on Guam to attack targets in the Western Pacific and on mainland Japan.

Walden Lee "Pug" Ainsworth (1886 – 1960) was an admiral of the United States Navy. For his role in commanding destroyer and cruiser task forces in the Pacific during the War, he was awarded the Navy Cross, the Navy Distinguished Service Medal, and the Legion of Merit.

W79 : Scott 493 : Marshall Islands 1994 : Warsaw uprising 1944.
Designer : Brian Sanders.
Selvedge commentary :
"Poles, the time of liberation is at hand!
Poles, to arms. There is not a moment to lose!" – Radio Kosciuszko, July 30, 1944.

The Warsaw Uprising was a major operation by the Polish underground resistance to liberate Warsaw from German occupation. It occurred in the summer of 1944, and it was led by the Polish resistance Home Army. The uprising was timed to coincide with the retreat of the German forces from Poland ahead of the Soviet advance. While approaching the eastern suburbs of the city, the Red Army halted combat operations, enabling the Germans to regroup and defeat the Polish resistance and to destroy the city in retaliation. The Uprising was fought for 63 days with little outside support. It was the single largest military effort taken by any European resistance movement during World War II.

The Uprising began on 1 August 1944 as part of a nationwide *Operation Tempest*, launched at the time of the Soviet Lublin–Brest Offensive. The main Polish objectives were to drive the Germans out of Warsaw while helping the Allies defeat Germany. An additional, political goal of the Polish Underground State was to liberate Poland's capital and assert Polish sovereignty before the Soviet-backed Polish Committee of National Liberation could assume control. Other immediate causes included a threat of mass German round-ups of able-bodied Poles for "evacuation"; calls by Radio Moscow's Polish Service for uprising; and an emotional Polish desire for justice and revenge against the enemy after five years of German occupation.

W80 : Scott 494 : Marshall Islands 1994 : The Liberation of Paris, 1944.
Designer : David K Stone.
Selvedge commentary :
" . . . fifteen solid miles of cheering, deliriously happy people waiting to shake your hand, to kiss you, to shower you with food and wine . . ." – Major Frank Burk, August 25, 1944.

The liberation of Paris was a battle that took place from 19 August 1944 until the German garrison surrendered the French capital on 25 August 1944. Paris had been occupied by Nazi Germany since the signing of the Armistice of 22 June 1940, after which the *Wehrmacht* occupied northern and western France.

The liberation began when the French Forces of the Interior—the military structure of the French Resistance—staged an uprising against the German garrison upon the approach of the US Third Army, led by General George S. Patton. On the night of 24 August, elements of General Philippe Leclerc de Hauteclocque's 2nd French Armoured Division made their way into Paris and arrived at the Hôtel de Ville shortly before midnight. The next morning, 25 August, the bulk of the 2nd Armoured Division and US 4th Infantry Division and other allied units entered the city. Dietrich von Choltitz, commander of the German garrison and the military governor of Paris, surrendered to the French at the Hôtel Le Meurice, the newly established French headquarters. General Charles de Gaulle of the French Army arrived to assume control of the city as head of the Provisional Government of the French Republic.

Philippe François Marie Leclerc de Hauteclocque (1902 – 1947) was a Free-French general during the war. He became Marshal of France posthumously in 1952, and is known in France simply as *le maréchal Leclerc* or just Leclerc. He fought in the Battle of France and was one of the first who defied his government's Armistice to make his way to Britain to fight with the Free French under General Charles de Gaulle, adopting the *nom de guerre* of Leclerc so that his wife and children would not be put at risk if his name appeared in the papers.

After the end of World War II in Europe in May 1945, he was given command of the French Far East Expeditionary Corps (*Corps expéditionnaire franÇais en ExtrêmeOrient*, CEFEO). He represented France at the surrender of the Japanese Empire in Tokyo Bay on 2 September 1945.

W81 : Scott 495 : Marshall Islands 1994 : U.S. Marines Land on Peleliu, 1944.
Designer : Brian Sanders.
Selvedge commentary :
" . . . *this is going to be a short one. Rough but fast. We'll be through in three days*"- Major
General William K Rupertus, September 15, 1944.

The vehicle shown in the image is an Armoured Amphibius Tractor LVT (A)-4.
The United States Army, Canadian Army and British Army used several LVT
models during the conflict, and referred to these vehicles as "Landing Vehicle,
Tracked."

The Battle of Peleliu, codenamed Operation Stalemate II by the US military, was
fought between the United States and Japan during the Mariana and Palau
Islands campaign of World War II, from 15 September to 27 November 1944,
on the island of Peleliu.

US Marines of the 1st Marine Division and then soldiers of the US Army's 81st
Infantry Division fought to capture an airfield on the small coral island of
Peleliu. The battle was part of a larger offensive campaign known as Operation
Forager, which ran from June to November 1944 in the Pacific Theatre.

Major General William Rupertus, the commander of the 1st Marine Division,
predicted that the island would be secured within four days. However, after
repeated Imperial Japanese Army defeats in previous island campaigns, Japan
had developed new island-defence tactics and well-crafted fortifications, which
allowed them to offer stiff resistance and extended the battle to more than two
months. The heavily outnumbered Japanese defenders put up such staunch
resistance, often fighting to the death in the name of the Japanese Emperor, that
the island became known in Japanese as the "Emperor's Island."

In the US, the battle was controversial because of the island's negligible strategic
value and the high casualty rate incurred by American troops during the fighting,

which exceeded that of all other amphibious operations during the Pacific War. The National Museum of the Marine Corps called it "the bitterest battle of the war for the Marines".

W82 : Scott 496 : Marshall Islands 1994 : MacArthur returns to the Philippines 1944.

Designer : David K Stone.

Selvedge commentary :

"People of the Philippines. I have returned! By the grace of Almighty God, our forces stand again on Philippines soil …"– General Douglas MacArthur, October 20, 1944.

The Philippines campaign, Battle of the Philippines, Second Philippines campaign, or the Liberation of the Philippines, codenamed Operation Musketeer I, II, and III, was the American, Filipino, Australian, and Mexican campaign to defeat and expel the Imperial Japanese forces occupying the Philippines.

The liberation of the Philippines from Japan commenced with amphibious landings on the eastern Philippine island of Leyte on October 20, 1944. The United States and Philippine Commonwealth military forces, with naval and air support from Australia and the Mexican 201st Fighter Squadron, were progressing in liberating territory and islands when the Japanese forces in the Philippines were ordered to surrender by Tokyo on August 15, 1945, after the dropping of the atomic bombs on mainland Japan and the Soviet invasion of Manchuria.

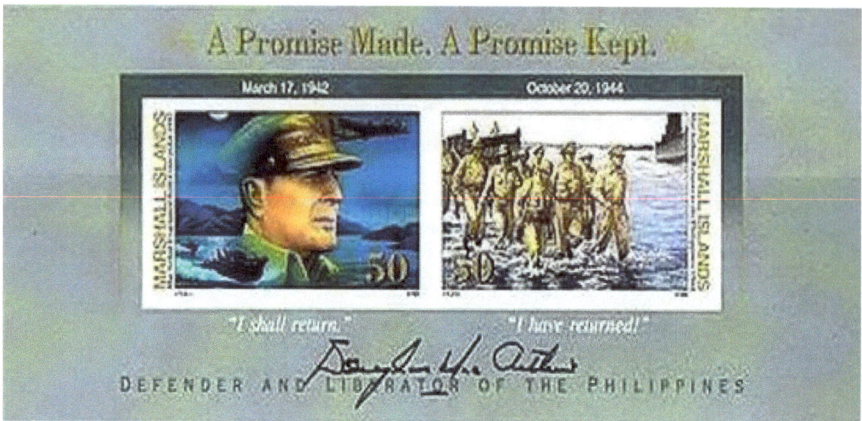

This miniature sheet, is extraneous to the "History of WW2" series :
The David Stone, Unicover painting is used in this issue; Scott 562 : Marshall
Islands 1994 : The 50th Anniversary of General Douglas MacArthur's Return to
Philippines, 1944.

W83 : Scott 497 : Marshall Islands 1994 : Battle of Leyte Gulf, 1944.
Designer : David K Stone.
Selvedge commentary :
"Strike! Repeat! Strike! Good luck!" – Admiral William F Halsey, October 24, 1944.

The Battle of Leyte Gulf, 23–26 October 1944, was the largest naval battle of World War II and by some criteria the largest naval battle in history, with over 200,000 naval personnel involved.

By the time of the battle, Japan had fewer capital ships (aircraft carriers and battleships) left than the Allied forces had total aircraft carriers in the Pacific, which underscored the disparity in force strength at that point in the war. Regardless, the IJN mobilized nearly all of its remaining major naval vessels in an attempt to defeat the Allied invasion of the Philippines, but it was repulsed by the US Navy's Third and Seventh Fleets.

The battle consisted of four main separate engagements (the Battle of the Sibuyan Sea, the Battle of Surigao Strait, the Battle off Cape Engaño, and the Battle off Samar), as well as lesser actions. Allied forces announced the end of organized Japanese resistance on the island of Leyte at the end of December.

It was the first battle in which Japanese aircraft carried out organized *kamikaze* attacks, and it was the last naval battle between battleships in history. The Japanese Navy suffered heavy losses and never sailed in comparable force thereafter, since it was stranded for lack of fuel in its bases, for the rest of the war.

W84 : Scott 498-499 : Marshall Islands 1994 : German Battleship Tirpitz sunk, 1944.

Designer : Brian Sanders.

Selvedge commentary :

"It is a great relief to get this brute where we have long wanted her"- Winston S Churchill, September 12, 1944.

The Brian Sanders images are of an *Avro Lancaster bomber* and the *German battleship Tirpitz*.

Tirpitz was the second of two *Bismarck*-class battleships built for Nazi Germany's *Kriegsmarine* (navy) prior to and during the Second World War. Named after Grand Admiral Alfred von Tirpitz, the architect of the *Kaiserliche Marine* (Imperial Navy), the ship was laid down at the *Kriegsmarinewerft* in Wilhelmshaven in November 1936 and her hull was launched two and a half years later. Work was completed in February 1941, when she was commissioned into the German fleet. Like her sister ship, *Bismarck*, *Tirpitz* was armed with a main battery of eight 38-centimetre (15 in) guns in four twin turrets. After a series of wartime modifications, she was 2000 tonnes heavier than *Bismarck*, making her the heaviest battleship ever built by a European navy.

After completing sea trials in early 1941, *Tirpitz* briefly served as the centrepiece of the Baltic Fleet, which was intended to prevent a possible break-out attempt by the Soviet Baltic Fleet. In early 1942, the ship sailed to Norway to act as a deterrent against an Allied invasion. While stationed in Norway, *Tirpitz* was also intended to be used to intercept Allied convoys to the Soviet Union, and two such missions were attempted in 1942. This was the only feasible role for her, since the St Nazaire Raid had made operations against the Atlantic convoy lanes too risky. *Tirpitz* acted as a fleet in being, forcing the British Royal Navy to retain significant naval forces in the area to contain the battleship.

In September 1943, *Tirpitz*, along with the battleship *Scharnhorst*, bombarded Allied positions on Spitzbergen, the only time the ship used her main battery in an offensive role. Shortly thereafter, the ship was damaged in an attack by British mini-submarines and subsequently subjected to a series of large-scale air raids. On 12 November 1944, British Lancaster bombers equipped with 12,000-pound (5,400 kg) "Tallboy" bombs scored two direct hits and a near miss which caused the ship to capsize rapidly. A deck fire spread to the ammunition magazine for one of the main battery turrets, which caused a large explosion. Figures for the number of men killed in the attack range from 950 to 1,204. Between 1948 and 1957, the wreck was broken up by a joint Norwegian and German salvage operation.

W85 : Scott 500-503 : Marshall Islands 1994 : The Battle of the Bulge, 1944.
Designer : David K Stone.
Selvedge commentaries :
"*Nuts*" – Brigadier General A McAuliffe, December 22, 1944.
"*We're going in now. Let 'er roll*" Lieutenant Colonel Creighton Abrams, December 26, 1944.

The two commentators are shown in (4) of the Stone illustration.

The Battle of the Bulge, also known as the Ardennes Offensive, was the last major German offensive campaign on the Western Front during World War II which took place from 16 December 1944 to 25 January 1945. It was launched through the densely forested Ardennes region between Belgium and Luxembourg.

The offensive was intended to stop Allied use of the Belgian port of Antwerp and to split the Allied lines, allowing the Germans to encircle and destroy each of the four Allied armies and force the western Allies to negotiate a peace treaty in the Axis powers' favour.

Allied forces eventually came to more than 700,000 men; from these there were from 77,000 to more than 83,000 casualties, including at least 8,600 killed. The

"Bulge" was the largest and bloodiest single battle fought by the United States in World War Two and the third-deadliest campaign in American history. It was one of the most important battles of the war, as it marked the last major offensive attempted by the Axis powers on the Western front. After this defeat, Nazi forces could only retreat for the remainder of the war.

W86 : Scott 504 : Marshall Islands 1995 : The Yalta Conference, 1945.
Designer : Shannon Steinweis.
Selvedge commentary :
"Our policy is not revenge; it is . . . to secure the future peace and safety of the world"-
Winston S Churchill

The Yalta Conference was held 4–11 February 1945, a meeting of the heads of
government of the United States, the United Kingdom and the Soviet Union to
discuss the postwar reorganization of Germany and Europe. The three states
were represented by President Franklin D. Roosevelt, Prime Minister Winston
Churchill, and General Secretary Joseph Stalin. The conference was held
near Yalta in Crimea, Soviet Union, within the Livadia, Yusupov,
and Vorontsov palaces.
The aim of the conference was to shape a postwar peace that represented not
only a collective security order, but also a plan to give self-determination to the
liberated peoples of Europe. Intended mainly to discuss the re-establishment of
the nations of war-torn Europe, within a few years, with the Cold War dividing
the continent, the conference became a subject of intense controversy.
During the Yalta Conference, the Western Allies had liberated all
of France and Belgium and were fighting on the western border of Germany. In
the east, Soviet forces were 40 miles from Berlin, having already pushed back the
Germans from Poland, Romania, and Bulgaria. There was no longer a question
regarding German defeat. The issue was the new shape of postwar Europe. The
French leader General Charles de Gaulle was not invited to either the Yalta
or Potsdam Conferences, a diplomatic slight that was the occasion for deep and
lasting resentment. De Gaulle attributed his exclusion from Yalta to the
longstanding personal antagonism towards him by Roosevelt, but the Soviets had
also objected to his inclusion as a full participant.

W87 : Scott 505 : Marshall Islands 1995 : Bombing of Dresden, 1945.
Designer : Brian Sanders.
Selvedge commentary :
"For the first time in many operations I felt sorry for the population below"- A British pilot,
February 1945.

The bombing of Dresden was a joint British and American aerial bombing attack
on the city of Dresden, the capital of the German state of Saxony, during World
War II. In four raids between 13 and 15 February 1945, 772 heavy bombers of
the Royal Air Force (RAF) and 527 of the United States Army Air
Forces (USAAF) dropped more than 3,900 tons of high-explosive bombs
and incendiary devices on the city. The bombing and the
resulting firestorm destroyed more than 1,600 acres of the city centre. Up to
25,000 people were killed. Three more USAAF air raids followed, two occurring
on 2 March aimed at the city's railway marshalling yard and one smaller raid on
17 April aimed at industrial areas.
Postwar discussions about whether the attacks were justified made the event a
moral *cause célèbre* of the war. Nazi Germany's desperate struggle to maintain
resistance in the closing months of the war is widely understood today, but
Allied intelligence assessments at the time painted a different picture. There was
uncertainty over whether the Soviets could sustain its advance on Germany, and
rumours of the establishment of a Nazi redoubt in Southern Germany were
taken too seriously.
The Allies saw the Dresden operation as the justified bombing of a strategic
target, which United States Air Force reports, declassified decades later, noted as
a major rail transport and communication centre, housing 110 factories and
50,000 workers supporting the German war effort. Several researchers later
asserted that not all communications infrastructure was targeted, and neither
were the extensive industrial areas located outside the city centre. Critics of the

bombing argue that Dresden was a cultural landmark with little strategic significance, and that the attacks were indiscriminate area bombing and were not proportionate to military gains. Some claim that the raid was a war crime. Nazi propaganda exaggerated the death toll of the bombing and its status as mass murder, and many in the German far-right have referred to it as "Dresden's Holocaust of bombs"█

---

[4] A personal comment : I had the privilege of working with an Australian Lancaster bomber pilot on his biography. His assertion was that he had been very well trained to carry out bombing missions from 20,000 feet. At the end of the war his British flight commander instructed Bill Gray to take his ground crew to view the effect of the bombing on Germany, Bill had achieved 29 missions. Dresden viewed from low-level was a revelation to him, it changed his whole perception of what he had been doing.

W88 : Scott 506 : Marshall Islands 1995 : Iwo Jima invaded by US Marines, 1945.
Designer : Howard Koslow.
Selvedge commentary :
"*Among the men who fought on Iwo Jima, uncommon valour was a common virtue*"- Admiral
Chester Nimitz, Mar16, 1945

The Battle of Iwo Jima (19 February – 26 March 1945) was a major battle in
which the United States Marine Corps and United States Navy landed on and
eventually captured the island of Iwo Jima from the Imperial Japanese
Army (IJA) during World War II. The American invasion, designated Operation
Detachment, had the purpose of capturing the island with its two airfields: South
Field and Central Field.
The Japanese Army positions on the island were heavily fortified, with a dense
network of bunkers, hidden artillery positions, and 18 km of tunnels. The
American ground forces were supported by extensive naval artillery and had
complete air supremacy provided by U.S. Navy and Marine Corps aviators
throughout the battle. The five-week battle saw some of the fiercest and
bloodiest fighting of the Pacific War.
Unique among Pacific War Marine battles, total American casualties exceeded
those of the Japanese, with a ratio of three American casualties for every two
Japanese. Of the 21,000 Japanese soldiers on Iwo Jima at the beginning of the
battle, only 216 were taken prisoner, some of whom were captured only because
they had been knocked unconscious or otherwise disabled. Most of the
remainder were killed in action, but it has been estimated that as many as 3,000
continued to resist within the various cave systems for many days afterwards
until they eventually succumbed to their injuries or surrendered weeks later.
The action was controversial, with retired Chief of Naval Operations William V.
Pratt stating that the island was useless to the Army as a staging base and useless
to the Navy as a fleet base. The Japanese continued to have early-warning radar

from Rota island, which was never invaded. Experiences with previous Pacific island battles suggested that the island would be well-defended and thus casualties would be significant. The lessons learned on Iwo Jima served as guidelines for the following Battle of Okinawa and the planned invasion of the Japanese homeland.

W89 : Scott 507 : Marshall Islands 1995 : Remagan Bridge taken by US Forces :
1945.

Designer : Shannon Stirnweis.

Selvedge commentary :

"This will bust 'em wide open. Shove everything you can across" – General Omar Bradley,
March 7, 1945.

The Battle of Remagen was an 18-day battle during the Allied invasion of
Germany in World War II. It lasted from 7 to 25 March 1945 when American
forces unexpectedly captured the Ludendorff Bridge over the Rhine intact. They
were able to hold it against German opposition and build additional temporary
crossings. The presence of a bridgehead across the Rhine advanced by three
weeks the Western Allies' planned crossing of the Rhine into the German
interior.

After capturing the Siegfried Line, the 9th Armoured Division of the U.S. First
Army had advanced unexpectedly quickly towards the Rhine. They were very
surprised to see one of the last bridges across the Rhine still standing. The
Germans had wired the bridge with about 2,800 kilograms (6,200 lb) of
demolition charges. When they tried to blow it up, only a portion of the
explosives detonated. U.S. forces captured the bridge and rapidly expanded their
first bridgehead across the Rhine, two weeks before Field Marshal Bernard
Montgomery's meticulously planned Operation Plunder. The U.S. Army's actions
prevented the Germans from regrouping east of the Rhine and consolidating
their positions.

The battle for control of the Ludendorff Bridge saw both the American and
German forces employ new weapons and tactics in combat for the first time.
Over the next 10 days, after the bridge's capture on 7 March 1945 and until its

failure on 17 March, the Germans used virtually every weapon at their disposal to try to destroy it. This included infantry and armour, howitzers, mortars, floating mines, mined boats, a railroad gun, and the 600 mm Karl-Gerät super-heavy mortar. They also attacked the bridge using the newly developed Arado Ar 234B-2 turbojet bombers. To protect the bridge against aircraft, the Americans positioned the largest concentration of anti-aircraft weapons during World War Two leading to "the greatest anti-aircraft artillery battles in American history". The Americans counted 367 different German Luftwaffe aircraft attacking the bridge over the next 10 days. The Americans claimed to have shot down nearly 30 percent of the aircraft dispatched against them. The German air offensive failed.

On 14 March, German Reich Chancellor Adolf Hitler ordered *Schutzstaffel* (SS) General Hans Kammler to fire V2 rockets to destroy the bridge. This marked the first time the missiles had been used against a tactical objective and the only time they were fired on a German target.

W90 : Scott 508 : Marshall Islands 1995 : Okinawa invaded by US Forces, 1945.
Designer : Shannon Stirnweis.
"Never before had I seen an invasion beach like Okinawa" – Reporter Ernie
Pyle, April 1, 1945.

The Battle of Okinawa codenamed *Operation Iceberg*, was a major battle of
the Pacific War fought on the island of Okinawa by United States
Army and United States Marine Corps forces against the Imperial Japanese
Army. The initial invasion of Okinawa on 1 April 1945 was the
largest amphibious assault in the Pacific Theatre of World War II. The Kerama
Islands surrounding Okinawa were pre-emptively captured on 26 March by
the 77th Infantry Division. The 82-day battle lasted from 1 April until 22 June
1945. After a long campaign of island hopping, the Allies were planning to
use Kadena Air Base on the large island of Okinawa as a base for *Operation
Downfall*, the planned invasion of the Japanese home islands, 340 miles away.

The battle has been referred to as the "typhoon of steel" The nickname refers to
the ferocity of the fighting, the intensity of Japanese *kamikaze* attacks and the
sheer numbers of Allied ships and armoured vehicles that assaulted the island.

The battle was the bloodiest and fiercest of the Pacific War, with some 50,000
Allied and around 100,000 Japanese casualties, also including
local Okinawans conscripted into the Japanese Army. According to local

authorities, at least 149,425 Okinawan people were killed, died by coerced suicide or went missing.

In the naval operations surrounding the battle, both sides lost considerable numbers of ships and aircraft, including the Japanese battleship *Yamato*. After the battle, Okinawa provided a fleet anchorage, troop staging areas, and airfields in proximity to Japan for US forces in preparation for a planned invasion of the Japanese home islands.

The *Yamato*-class battleships were two battleships of the Imperial Japanese Navy, *Yamato* and *Musashi*, laid down leading up to World War II and completed as designed. A third hull, laid down in 1940, was converted to an aircraft carrier, *Shinano*, during construction.

Displacing nearly 72,000 long tons at full load, the completed battleships were the heaviest ever constructed in the world. Due to the threat of U.S. submarines and aircraft carriers, both *Yamato* and *Musashi* spent the majority of their careers in naval bases at Brunei, Truk, and Kure—deploying on several occasions in response to U.S. raids on Japanese bases.

All three ships were sunk by the U.S. Navy; *Yamato* was sunk by air strikes while en- route from Japan to Okinawa as part of *Operation Ten-Go* in April 1945.

W91 : Scott 509 : Marshall Islands 1995 : Death of Franklin D Roosevelt, 1945.
Designer : Howard Koslow.
Selvedge commentary :
"*ARMY - NAVY DEAD : Roosevelt, Franklin, D, Commander in Chief*". Official
Bulletin, April 12, 1945.

Roosevelt's health was in decline as he prepared in 1944 for both a fourth run at
the presidency and the aftermath of World War II. A March 1944 examination
by his doctors revealed a variety of heart ailments, high blood pressure, and
bronchitis. Those close to the President—and even those who saw him speak in
public—noted his haggard and weak appearance, his flagging energy, and his
increasing lapses of concentration and memory. Most of the American public
was unaware of the President's struggles—though rumours about FDR's health
often ran wild—and FDR delivered a few key, command performances in 1944
that quieted concerns. Nonetheless, Roosevelt's election victory over Thomas E.
Dewey in 1944, in addition to the Yalta Conference the following February, put
the President under immense strain. In April 1945, FDR returned to Warm
Springs, Georgia, a destination that had served since the 1920s as his favourite
retreat. There, on April 12, while sitting for a portrait, he collapsed and died of a
cerebral haemorrhage. Vice President Harry Truman took the oath of office the
same day.

W92 : Scott 510 : Marshall Islands 1995 : US / USSR troops link at Elbe River, 1945.

Designer: Gherman Comlev.

Selvedge commentary :

"This is an historic occasion. It is a moment for which both our armies have been fighting"- Soviet Lieutenant, April 25, 1945.

Elbe Day, April 25, 1945, is the day Soviet and American troops met at the Elbe River, near Torgau in Germany, marking an important step toward the end of World War II in Europe. This contact between the Soviets, advancing from the east, and the Americans, advancing from the west, meant that the two powers had effectively cut Germany in two.

W93 : Scott 511 : Marshall Islands 1995 : Soviet troops conquer Berlin, 1945.
Designer : Ivan Sushchenko.
Selvedge commentary :
"Fascist Germany has been forced to her knees by the Red Army and the forces of our Allies"-
Joseph Stalin, May 9, 1945.

The Battle of Berlin, designated as the Berlin Strategic Offensive Operation by
the Soviet Union, and also known as the Fall of Berlin, was one of the last
major offensives of the European theatre of World War II.

After the Vistula–Oder offensive of January–February 1945, the Red Army had
temporarily halted on a line 60 km (37 mi) east of Berlin. On 9 March, Germany
established its defence plan for the city with Operation Clausewitz. The first
defensive preparations at the outskirts of Berlin were made on 20 March, under
the newly appointed commander of Army Group Vistula, General Gotthard
Heinrici.

When the Soviet offensive resumed on 16 April, two Soviet fronts (army groups)
attacked Berlin from the east and south, while a third overran German forces
positioned north of Berlin. Before the main battle in Berlin commenced, the Red
Army encircled the city after successful battles of the Seelow Heights and Halbe.
On 20 April 1945, Hitler's birthday, the 1st Belorussian Front led
by Marshal Georgy Zhukov, advancing from the east and north, started shelling
Berlin's city centre, while Marshal Ivan Konev's 1st Ukrainian Front broke
through Army Group Centre and advanced towards the southern suburbs of
Berlin. On 23 April General Helmuth Weidling assumed command of the forces
within Berlin. The garrison consisted of several depleted and
disorganised Army and Waffen-SS divisions, along with poorly
trained *Volkssturm* and Hitler Youth members. Over the course of the next week,
the Red Army gradually took the entire city.

On 30 April, Hitler committed suicide. The city's garrison surrendered on 2 May but fighting continued to the north-west, west, and south-west of the city until the end of the war in Europe on 8 May (9 May in the Soviet Union) as some German units fought westward so that they could surrender to the Western Allies rather than to the Soviets.

During and immediately following the assault, in many areas of the city, vengeful Soviet troops (often rear echelon units) engaged in mass rape, pillage and murder. Oleg Budnitskii, historian at the Higher School of Economics in Moscow, told a BBC Radio programme that Red Army soldiers were astounded when they reached Germany. "For the first time in their lives, eight million Soviet people came abroad, the Soviet Union was a closed country. All they knew about foreign countries was there was unemployment, starvation and exploitation. And when they came to Europe they saw something very different from Stalinist Russia ... especially Germany. They were really furious, they could not understand why being so rich, Germans came to Russia".Other authors question the narrative of sexual violence by Red Army soldiers being more than what was a sad normality from all sides during the war, including the Western Allies. Nikolai Berzarin, commander of the Red Army in Berlin, quickly introduced penalties up to the death penalty for looting and rape.

Despite Soviet efforts to supply food and rebuild the city, starvation remained a problem.

W94 : Scott 512 : Marshall Islands 1995 : Allies liberate Concentration Camps, 1945.

Designer : Howard Koslow.

Selvedge Commentary :

" . . . they were waiting for somebody to turn their lives back straight again . . . they were just lost souls" – John Glustrom, US 333rd Engineers.

As Allied troops moved across Europe in a series of offensives against Germany, they began to encounter and liberate concentration camp prisoners, many of whom had survived death marches into the interior of Germany. Soviet forces were the first to approach a major Nazi camp, reaching the Majdanek camp near Lublin, Poland, in July 1944. Surprised by the rapid Soviet advance, the Germans attempted to demolish the camp in an effort to hide the evidence of mass murder.

The Soviets also liberated major Nazi camps,
including Auschwitz, Stutthof, Sachsenhausen, and Ravensbrück.
US forces liberated the Buchenwald, Dora-Mittelbau, Flossenbürg, Dachau, and Mauthausen camps.
British forces liberated camps in northern Germany,
including Neuengamme and Bergen-Belsen.

Liberators confronted unspeakable conditions in the Nazi camps, where piles of corpses lay unburied. Only after the liberation of these camps was the full scope of Nazi horrors exposed to the world. The small percentage of inmates who survived resembled skeletons because of the demands of forced labour and the lack of food, compounded by months and years of maltreatment. Many were so weak that they could hardly move. Disease remained an ever-present danger, and many of the camps had to be burned down to prevent the spread of epidemics. Survivors of the camps faced a long and difficult road to recovery. (*US Holocaust Memorial Museum*).

W95 : Scott 513-516 : Marshall Islands 1995 : VE Day – Victory in Europe, 1945.   See also page 204.

Selvedge commentary :

"We the undersigned, acting by authority of the German High Command, hereby surrender unconditionally … all forces on land, sea and in the air" – Colonel General Alfred Jodl, Rheims 02.41 hours, May 7, 1945.

"The Allied Armies, through sacrifice and devotion have wrung from Germany a final and unconditional surrender" – President Harry S Truman, Washington, May 8, 1945.

The separate images are not titled.

Upon the defeat of Germany, celebrations erupted throughout the Western world, especially in the United Kingdom, in North America and in USSR. More than one million people celebrated in the streets throughout the UK to mark the end of the European part of the war. In London, crowds massed in Trafalgar Square and up the Mall to Buckingham Palace, where King George VI and Queen Elizabeth, accompanied by their daughters and Prime Minister Winston Churchill, appeared on the balcony of the palace before the cheering crowds. Churchill went from the palace to Whitehall, where he addressed another large crowd:

"God bless you all. This is your victory. In our long history, we have never seen a greater day than this. Everyone, man or woman, has done their best".

Churchill asked Ernest Bevin to come forward and share the applause. Bevin said, "No, Winston, this is your day", and proceeded to conduct the people in the singing of "For He's a Jolly Good Fellow".

In the United States, the event coincided with President Harry S. Truman's 61st birthday. He dedicated the victory to the memory of his predecessor, Franklin D. Roosevelt, who had died of a cerebral haemorrhage less than a month earlier, on 12 April. Flags remained at half-staff for the remainder of the 30-day mourning period. Truman said of dedicating the victory to Roosevelt's memory and keeping the flags at half-staff that his only wish was "that Franklin D. Roosevelt had lived to witness this day". Later that day, Truman said that the victory made it his most enjoyable birthday. Great celebrations took place in many American cities, especially in New York's Times Square.

Tempering the jubilation somewhat, both Churchill and Truman pointed out that the war against Japan had not yet been won. In his radio broadcast at 15:00 on 8 May, Churchill told the British people, "We may allow ourselves a brief period of rejoicing as Japan remains unsubdued". In America, Truman broadcast at 09:00 and said it was "a victory only half won".

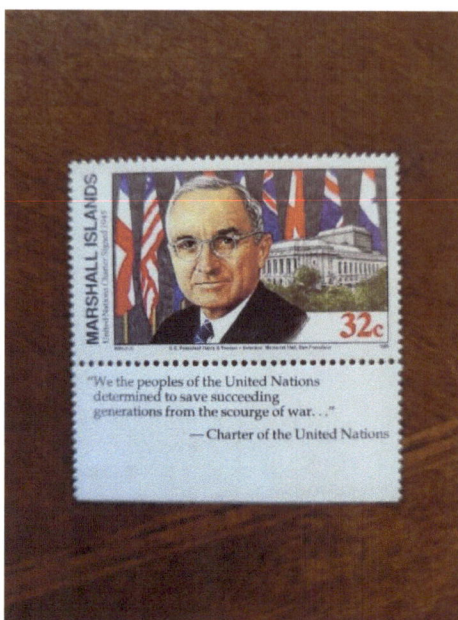

W96 : Scott 517 : Marshall Islands 1995 : United Nations Charter, 1945.
Designer : Brian Sanders.
Selvedge commentary :
"We the peoples of the United Nations determined to save succeeding generations from the scourge of war ..."- Charter of the United Nations.

During the Second World War, the Allies agreed to establish a new postwar international organization. The Charter of the United Nations (UN) is the foundational treaty of the United Nations. It establishes the purposes, governing structure, and overall framework of the UN system, including its six principal organs: the Secretariat, the General Assembly, the Security Council, the Economic and Social Council, the International Court of Justice, and the Trusteeship Council.

The UN Charter mandates the UN and its member states to maintain international peace and security, uphold international law, achieve "higher standards of living" for their citizens, address "economic, social, health, and related problems", and promote "universal respect for, and observance of, human rights and fundamental freedoms for all without distinction as to race, sex, language, or religion". As a charter and constituent treaty, its rules and obligations are binding on all members and supersede those of other treaties.

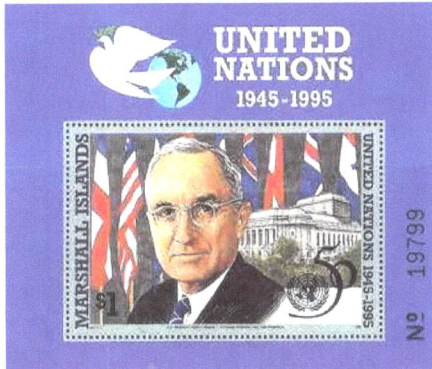

A second image, extraneous to the "History of WW2" series : Scott 563 souvenir sheet : Marshall Islands 1995 : United Nations Charter, 1945 and President Harry Truman. Issued at the same time as the image within the series.

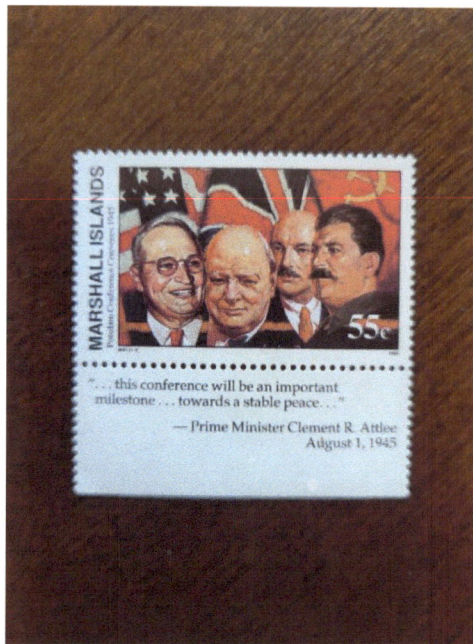

W97 : Scott 518 : Marshall Islands 1995 : Potsdam Conference convenes, 1945.
Designer : Shannon Stirnweis.
Selvedge commentary :
"... this conference will be an important milestone... towards a stable peace"- Prime
Minister Clement Atlee (Great Britain), August 1, 1945.

The Potsdam Conference was held at Potsdam in the Soviet occupation zone
from July 17 to August 2, 1945, to allow the three leading Allies to plan the
postwar peace, while avoiding the mistakes of the Paris Peace Conference of
1919. The participants were the Soviet Union, the United Kingdom, and the
United States. They were represented respectively by General Secretary Joseph
Stalin, Prime Ministers Winston Churchill and Clement Attlee, and
President Harry S. Truman. They gathered to decide how to administer
Germany, which had agreed to an unconditional surrender nine weeks earlier.
The goals of the conference also included establishing the postwar order, solving
issues on the peace treaty, and countering the effects of the war.

The foreign ministers and aides played key roles: Vyacheslav Molotov, Anthony
Eden and Ernest Bevin, and James F. Byrnes. From July 17 to July 25, nine
meetings were held, when the Conference was interrupted for two days, as the

results of the British general election were announced. By July 28, Attlee had defeated Churchill and replaced him as Britain's representative, with Britain's new Secretary of State for Foreign Affairs, Ernest Bevin, replacing Anthony Eden.

Key final decisions included the following: Germany would be divided into the four occupation zones (among the three powers and France) that had been agreed to earlier; Germany's eastern border was to be shifted west to the Oder–Neisse line; a Soviet-backed group was recognized as the legitimate government of Poland; and Vietnam was to be partitioned at the 16th parallel. The Soviets also reaffirmed their Yalta promise to promptly launch an invasion of Japanese-held areas.

W98 : Scott 519 : Marshall Islands 1995 : The resignation of Winston Churchill, 1945.
Designer : Brian Sanders.
Selvedge commentary : ". . . the vote of the nation at the general election was one of the greatest disasters . . . in our . . . history"- Winston S Churchill.

I am not sure that 'resignation' reflects the reality of Churchill losing Office when his Conservative Party was defeated by the UK Labour Party in a General Election towards the end of the war. He lost his position as Prime Minister. The Churchill war ministry was the United Kingdom's coalition government for most of the Second World War from 10 May 1940 to 23 May 1945. It was led by Winston Churchill, who was appointed prime minister of the United Kingdom by King George VI following the resignation of Neville Chamberlain in the aftermath of the Norway Debate.

At the outset, Churchill formed a five-man war cabinet which included Chamberlain as Lord President of the Council, Clement Attlee as Lord Privy Seal and later as Deputy Prime Minister, Viscount Halifax as Foreign Secretary, and Arthur Greenwood as a minister without portfolio. Although the original war cabinet was limited to five members, in practice they were augmented by the service chiefs and ministers who attended the majority of meetings. The cabinet changed in size and membership as the war progressed but there were significant additions later in 1940 when it was increased to eight after Churchill, Attlee, and Greenwood were joined by Ernest Bevin as Minister of Labour and National Service; Anthony Eden as Foreign Secretary, replacing Halifax who was sent to Washington as Ambassador to the United States; Lord Beaverbrook as Minister of Aircraft Production; Sir Kingsley Wood as Chancellor of the Exchequer; and Sir John Anderson as Lord President of the Council, replacing Chamberlain who died in November (Anderson later became chancellor after Wood's death in September 1943).

The coalition was dissolved in May 1945, following the final defeat of Germany, when the Labour Party decided to withdraw in order to prepare for a general election. Churchill, who was the leader of the Conservative Party, was asked by the King to form a new, essentially Conservative, government. It was known as the Churchill caretaker ministry and managed the country's affairs until the completion of the general election on 26 July that year.

W99 : Scott 520 Marshall Islands 1995 : Atomic bomb dropped on Hiroshima, 1945.
Designer : Nikolai Litvinov.
Selvedge commentary :
"We must bow to the inevitable . . . we must put an end to the war as soon as possible".
Emperor Hirohita, August 6, 1945.

The B-29 aircraft represented within the stamp image, by Nikolai Litvinov, is named the Enola Gay.

On 6 and 9 August 1945, the United States detonated two atomic bombs over the Japanese cities of Hiroshima and Nagasaki. The bombings killed between 150,000 and 246,000 people, most of whom were civilians, and remain the only use of nuclear weapons in an armed conflict. Japan surrendered to the Allies on 15 August, six days after the bombing of Nagasaki and the Soviet Union's declaration of war against Japan and invasion of Japanese-occupied Manchuria. The Japanese government signed the instrument of surrender on 2 September, effectively ending the war.
In the final year of World War II, the Allies prepared for a costly invasion of the Japanese mainland. This undertaking was preceded by a conventional bombing and firebombing campaign that devastated 64 Japanese cities. The war in the European theatre concluded when Germany surrendered on 8 May 1945, and the Allies turned their full attention to the Pacific War. By July 1945, the Allies' Manhattan Project had produced two types of atomic bombs: "Little Boy", an enriched uranium gun-type fission weapon, and "Fat Man", a plutonium implosion-type nuclear weapon. The 509th Composite Group of the United States Army Air Forces was trained and equipped with the specialized Silverplate version of the Boeing B-29 Superfortress, and deployed to Tinian in the Mariana Islands. The Allies called for the unconditional surrender of the Imperial Japanese armed forces in the Potsdam Declaration on 26 July 1945, the alternative being "prompt and utter destruction". The Japanese government ignored the ultimatum.

The consent of the United Kingdom was obtained for the bombing, as was required by the Quebec Agreement, and orders were issued on 25 July by General Thomas Handy, the acting chief of staff of the United States Army, for atomic bombs to be used against Hiroshima, Kokura, Niigata, and Nagasaki. These targets were chosen because they were large urban areas that also held militarily significant facilities. On 6 August, a Little Boy was dropped on Hiroshima. Three days later, a Fat Man was dropped on Nagasaki. Over the next two to four months, the effects of the atomic bombings killed 90,000 to 166,000 people in Hiroshima and 60,000 to 80,000 people in Nagasaki; roughly half occurred on the first day. For months afterward, many people continued to die from the effects of burns, radiation sickness, and other injuries, compounded by illness and malnutrition. Though Hiroshima had a sizable military garrison, most of the dead were civilians.

Scholars have extensively studied the effects of the bombings on the social and political character of subsequent world history and popular culture, and there is still much debate concerning the ethical and legal justification for the bombings. According to supporters, the atomic bombings were necessary to bring an end to the war with minimal casualties and ultimately prevented a greater loss of life; according to critics, the bombings were unnecessary for the war's end and were a war crime, raising moral and ethical implications.

W100 : Scott 521-524 : Marshall Islands 1995 : V-J Day, 'Victory over Japan day',
1945.   See also page 205.

Designer : Brian Sanders.

Selvedge commentary :

"The issues . . . have been determined on the Battlefields . . ."

"Let us pray that peace is now restored to the world . . . " - General Douglas MacArthur,
September 2, 1945.

The separate images are not titled.

The surrender of the Empire of Japan in World War II was announced by
Emperor Hirohito on 15 August and formally signed on 2 September
1945, ending the war. By the end of July 1945, the Imperial Japanese Navy (IJN)
was incapable of conducting major operations and an Allied invasion of
Japan was imminent. Together with the United Kingdom and China, the United
States called for the unconditional surrender of Japan in the Potsdam
Declaration on 26 July 1945—the alternative being "prompt and utter
destruction". While publicly stating their intent to fight on to the bitter end,
Japan's leaders (the Supreme Council for the Direction of the War, also known
as the "Big Six") were privately making entreaties to the publicly neutral Soviet
Union to mediate peace on terms more favourable to the Japanese. While
maintaining a sufficient level of diplomatic engagement with the Japanese to give
them the impression they might be willing to mediate, the Soviets were covertly
preparing to attack Japanese forces in Manchuria and Korea (in addition

to South Sakhalin and the Kuril Islands) in fulfillment of promises they had secretly made to the US and the UK at the Tehran and Yalta Conferences.

On 28 August, the occupation of Japan led by the Supreme Commander for the Allied Powers began. The surrender ceremony was held on 2 September, aboard the United States Navy battleship USS *Missouri*, at which officials from the Japanese government signed the Japanese Instrument of Surrender, ending the hostilities. Allied civilians and military personnel alike celebrated V-J Day, the end of the war; however, isolated soldiers and personnel from Japan's forces throughout Asia and the Pacific refused to surrender for months and years afterwards, some into the 1970s. The role of the atomic bombings in Japan's unconditional surrender, and the ethics of the two attacks, is debated. The state of war formally ended when the Treaty of San Francisco came into force on 28 April 1952. Four more years passed before Japan and the Soviet Union signed the Soviet–Japanese Joint Declaration of 1956, which formally brought an end to their state of war.

Designed, as if it is a legitimate part of the "History of WW2", I have determined to view the next image as being outside the scope of this study.

Marshall Islands 1990 : The re-unification of Germany 1990.

The East German government dominated by the Socialist Unity Party of Germany, (a communist party), started to falter on 2 May 1989, when the removal of Hungary's border fence with Austria opened a hole in the Iron Curtain. The border was still closely guarded, but the Pan-European Picnic and the indecisive reaction of the rulers of the Eastern Bloc set in motion an irreversible movement. It allowed an exodus of thousands of East Germans fleeing to West Germany via Hungary. The Peaceful Revolution, a part of the international Revolutions of 1989 including a series of protests by the East German citizens, led to the fall of the Berlin Wall on 9 November 1989 and GDR's first free elections later on 18 March 1990 and then to the negotiations between the two countries that culminated in a Unification Treaty.

German reunification was the process of re-establishing Germany as a single full sovereign state, which took place between 9 November 1989 and 15 March 1991. The "Unification Treaty" entered into force on 3 October 1990, dissolving the German Democratic Republic, DDR, or East Germany and integrating its recently re-established constituent federated states into the Federal Republic of Germany or West Germany to form present-day Germany.

An example of the imagination shown by one E-Bay vendor to illustrate the stamps they have for sale. This vendor invites the purchaser to choose two examples (of the blocks of four stamps) at the discounted price of US$2.10 each.

Taking the WW2 series images into other commemorations.

Quite late in the study I came across the fact that Unicover had re-used three of the blocks of four images of the main 1989-1995 set as commemorative images for subsequent issues. These are shown below:

C221: Scott 834 : Marshall Islands 2004. 60th Anniversary of D-Day,
May 6, 2004. See also page 154.

The recognised equipments are : (1) Horsa Gliders and Parachute Troops, (2) British Typhoon-1B and US P51B Mustangs, (3) German Defences and Pointe du Hoc and (4) Allied Amphibious Landing. An additional feature is that each image carries its Unicover number and a copyright date, for example the third image ©includes its identity C221.(4-3) ©2004.

I have been particularly interested in these images as Unicover reprinted them. They were still working directly with The Marshall Islands Postal Services. Discussing this fact, via the Internet with Brookman Stamps, they advised me that they had always believed that the reprints were to correct factual errors within the images or descriptions. Within this block of four stamp (3) "German Defences and Pointe du Hoc". I have looked up the use or not of the Hawker Hurricane during D-Day, in Britannica. They were not used

Two of the Marshall Island series reuse have prompted a French First Day Cover specialist to produce good-looking covers. Historiaphil have the address of *Historiaphil © Kiosque de Philatélie du Parlement Européen, Strasbourg* These two covers are available for €20 and I believe they enhance the blocks of four stamps in both cases.

C236 : Scott 857 : Marshall Islands 2005. 60th Anniversary of VE-Day.
May 8, 2005.   See also page

The images are entitled (1) German Surrender, Rheims, (2) Times Square, New York, (3) Victory Parade, Moscow, (4) Buckingham Palace, London. The selvedge shows the copyright assertion ©2005 MIPS (Marshall Islands Postal Service). The individual stamps show their position in the block of four stamps, for example the third image carries the legend C236. (4-3) ©2005.

C239 : Scott 866 : Marshall Islands 2005. 60th Anniversary of VJ-Day.
September 2, 2005.   See also page

The images are entitled (1) Fujiama and Tokyo Bay, (2) USS Missouri, (3) US
Contingent, (4) Japanese Delegation. The selvedge shows the copyright assertion
©2005 MIPS (Marshall Islands Postal Service). The individual stamps show their
position in the block of four stamps, for example the third image carries the
legend C239. (4-3) ©2005.

Additional notes :

The Trust Territory of the Pacific Islands (TTPI) was a United Nations trust territory in Micronesia administered by the United States from 1947 to 1994. The Imperial Japanese South Seas Mandate had been seized by the US during the Pacific War, as Japan had administered the territory since the League of Nations gave Japan mandate over the area from Imperial Germany after World War I. However, in the 1930s, Japan left the League of Nations and then invaded additional lands. During World War II, military control of the islands was disputed, but by the end of the war the islands had come under control of the Allies. The Trust Territory of the Pacific was created to administer the islands as part of the United States, while still under the auspices of the United Nations. Most of the island groups in the territory became independent states, with some degree of ties kept with the United States: the Federated States of Micronesia, Marshall Islands and Palau are today independent states in a Compact of Free Association with the US, while the Northern Mariana Islands remain under US jurisdiction, as an unincorporated territory and commonwealth.

The Marshall Islands (the Federated Staes of Micronesia and Palau) have as sovereign states have become freely associated with the United States under the Compact of Free Association (COFA).

Sources for the stamp images I have reproduced in this review

I have not been able to use my own collection very much as prior to my writing this project I was largely unaware of the Marshall Islands. As a collector I had chosen to only collect, when available, stamps that have fulfilled their niche role and have been used to carry mail. The postal authority has officially cancelled these stamps with, hopefully, a circular date-stamp showing the entry point into the system and that date of posting. Occasionally this 'spoils' the image, but not often. I have chosen to use the sharpest image that I have when scanning my own stamps.

For this study I have been blessed with the direct help of former Unicover / Fleetwood staff to receive stamps from their own resources.

I have made extensive use of stamp catalogues to garner appropriate representations of battles / conflicts and confirm I am not missing any images. Catalogues come in printed form and / or in digital form via the Internet.

On a world-wide basis four catalogues are used as 'bibles', these are :

1. Michel: Deutschland, Europa and Überseekatalog
2. Scott: Standard Postage Stamp Catalog
3. Stanley Gibbons: Stamps of the World
4. Yvert & Tellier: Catalogue de Timbres-Poste

Where necessary I have previously used the digital images of the *Stampworld* catalogue, the Postal Authorities own websites and commercial sales outlets, such as *e-bay*, to find the appropriate image I was looking for.

Historical integrity

Defining their integrity Feetwood created a Military Historical Advisory Board to select the 100 epic events of World War II. The brief was that the stamp images should capture the tragedy and futility of war with authentic designs, thereby making a dramatic and enduring peace statement. At the same time the stamp images are a lasting tribute to human courage and the quest for freedom. The panel was convened 50 years after the war and the history of the time determined the contexts of the events from a largely American perspective – and no problem with that.

Major General John W. Huston
U.S. Air Force, Retired

After flying B-17 bombers during World War II, Huston served as Chairman of the History Department at the United States Naval Academy, and as Chief of the Office of Air Force History in Washington, D.C. He is the author of numerous scholarly articles on aviation history, and Chairman of the Military Historical Advisory Board.

Brigadier General John S.D. Eisenhower
Army of the U.S., Retired

After graduating from the U.S. Military Academy at West Point, Eisenhower served with the 1st U.S. Army in World War II, then undertook a variety of assignments before being appointed U.S. ambassador to Belgium in 1969. His many books include *The Bitter Woods*, *Letters to Mamie* and *So Far From God*.

General Sir Anthony Farrar-Hockley
British Army, Retired

After earning a battlefield commission during World War II, Farrar-Hockley went on to hold a number of prestigious posts, among them instructor at the Royal Military College and commander in chief of Allied Forces Northern Europe (NATO). His many books include *War in the Desert*.

Edward C. Ezell, Ph.D.
Division of Armed Forces History, Smithsonian Institution

The author of *Personal Firepower*, a history of the Vietnam War, Ezell earned his Ph.D. from the Case Institute of Technology in Cleveland, Ohio. A long-time military enthusiast, he presently serves as curator of the Division of Armed Forces History at the Smithsonian Institution in Washington, D.C.

Vice Admiral William P. Mack
U.S. Navy, Retired

A graduate of the U.S. Naval Academy, Mack was commander of destroyers in the Pacific during World War II, then held a variety of peacetime assignments — including commander of the U.S. 7th Fleet, assistant secretary of defense, and superintendent of the U.S. Naval Academy. He is co-author of *South to Java*.

Major General Walter H. Baxter III
U.S. Air Force, Retired

After serving in World War II, Baxter graduated from West Point and was a pilot in Korea and Vietnam. His many air command duties included Northern European NATO Command and the 24th North American Air Defense Command. He has also written several military texts.

I have looked on the Internet to confirm the bona fides of the panel, the choice of Fleetwood, and who are we to question their integrity?

The stamp designers

Fleetwood commissioned 9 artists to illustrate the epic events of the second World War. They are recognised in the Project History.

David K Stone (1922-2001), who contributed 37 of the 157 images used within the set.

	David K Stone was a member of the US Defence Forces from 1943 to 1946. He served as a First lieutenant with the occupation forces in Japan Upon discharge he made his way to New York and a career as an illustrator artist. A search of the Internet shows examples of his paintings, some of which found their way into the Marshall Islands set.

Examples of David K Stone's paintings, the Liberation of Paris and a Vietnam detail, downloaded from the Internet during December 2024.

Brian Sanders (born 1937), who contributed 81 of the 157 images used within the Fleetwood set.

Brian Sanders' career in advertising was interrupted in 1956 by National Service within the UK defence system. Serving in the Royal Marines he saw action during the Suez crisis, spending most of his service in and around North Africa and the Mediterranean. In the final year of his service he worked in surveillance, largely due to his technical drawing skills.

In 1958 he became a free-lance career as a graphic designer. In 1959 Sanders joined Artist Partners, a leading London Creative agency set up in 1950. Based in Soho, it represented leading artists, illustrators, designers and photographers. Amongst Sanders' clients is Royal Mail. Sanders continues to work today and remains a respected figure in the illustration world; through the sheer body of work he has completed during his extensive career, academic recognition due to his work as an assessor for various art schools and his professional contribution to the industry as a founding member of the Association of Illustrators.

Brian Sanders

The first Battle of Savo Island and a non-titled painting.

The other artists who contributed are named and photographed in the Fleetwood book. They include Chris and Paul Calle, (3) Gherman Komlev (-) Howard Koslow (16), Shannon Stirnweis (13), Ivan Sushchenko (1) and Nikolai Litvinov (1). It has been difficult to find updated biographies of these artists. The Internet concentrates on the availability of their work.

When is a stamp NOT a stamp?

With the advent of the Internet the world has embraced the digital world. Written and typed postage as the main means of communications between the public and business has reduced by 85%. In an attempt to retain revenues, the postal authorities have resorted to printing more stamps in sheer numbers, raising the cost of postage year-on year and introducing nefarious reasons for distributing new issues. The target markets for the postal authorities now include the casual buyer, theme collectors and souvenir hunters and the 15% historical die-hards. Some countries have adopted an over-the-top policy and until Covid the industry sensed that the numbers of serious collectors were diminishing to the benefit of the specialist stamp auctioneers. Covid provided, however, a respite and stamp collecting became a fashionable hobby again – but even more expensive.

Four independent stamp agencies, plus a few other organisations, on contract take care of the designing, manufacturing, issuing, marketing and selling of postage stamps on behalf of a country's postal administration. Through my previous studies I have come across the activities of :

> The Inter-Governmental Philatelic Corporation (IGPC) of New York,
> Impressor SA based in Syens, Switzerland,
> Unicover Corporation of Wyoming, (now trading as the Mystic Stamp Company).
> Stamperija Limited of Vilnius, Lithuania.

These companies have a style that is their own and recognisable to the collector. For each subject selected to provide the excuse to print a set of stamps they will generally issue four stamps, (within a miniature sheet) and a higher value stamp also within a miniature sheet. The subjects offered are those that the design-agency believes will sell the most copies. These pseudo-stamps look like stamps and are available to the collector, not necessarily from the country's post offices but commercially, and not necessarily for the implied service-fee. However :

> The Universal Postal Union (UPU) in 2016 published a revised
> Philatelic Code of Ethics for the use of UPU members when
> issuing and supplying postage stamps and other philatelic products.

> According to it when choosing themes and other design elements
> the issuing postal authorities *shall not produce postage stamps or philatelic*

products that are intended to exploit customers. In other words issuing postal authorities shall not produce any abusive issues. (www.fitphitalelicphilately.org/2020/TCNewsNo29 September2020).

There is an argument to suggest that the Unicover supply of the set, we are looking at, might be deemed to fall within the definition of abusive issues. The level of integrity and relevance of the stamp images and presentation removes that risk. The series is as potent and vital as ever.

So the Fleetwoord set is the best thematic set ever ....

I believe so. Who else was in the evaluation?

The Centenary of the First World War, during 2014-2018 prompted many postal authorities to consider special and part-publishing sets over a five year period. New Zealand Post produced a box-set of five booklets designed on a first World War soldiers' diary format, pages bound together by string. The issues were entitled :

     1914 For King and Country
     1915 The spirit of ANZAC
     1916 Courage and Commitment
     1917 The darkest hour
     1918 Back from the brink

Every stamp contains a single red-poppy image. The stamps have been issued as singles, as a set of ten stamps within a miniature sheet, grouped together in two other miniature sheets, and contained in the prestige stamp book.

This illustration shows the (diary) Prestige book cover and the miniature sheet set for the 2014 issue.

The two third best sets, to my mind are the Unicover / Fleetwood 70th Anniversary commemorative set of 2011 built upon the history described above and the United States Postal Service fiftieth year anniversity of the end of WW2 set, published at the same time as the RMI set. Images from these issues are reproduced below.

Chris Calle was a contribing artist to the main set we are discussing.

He has also created the images for the set reproduced here.

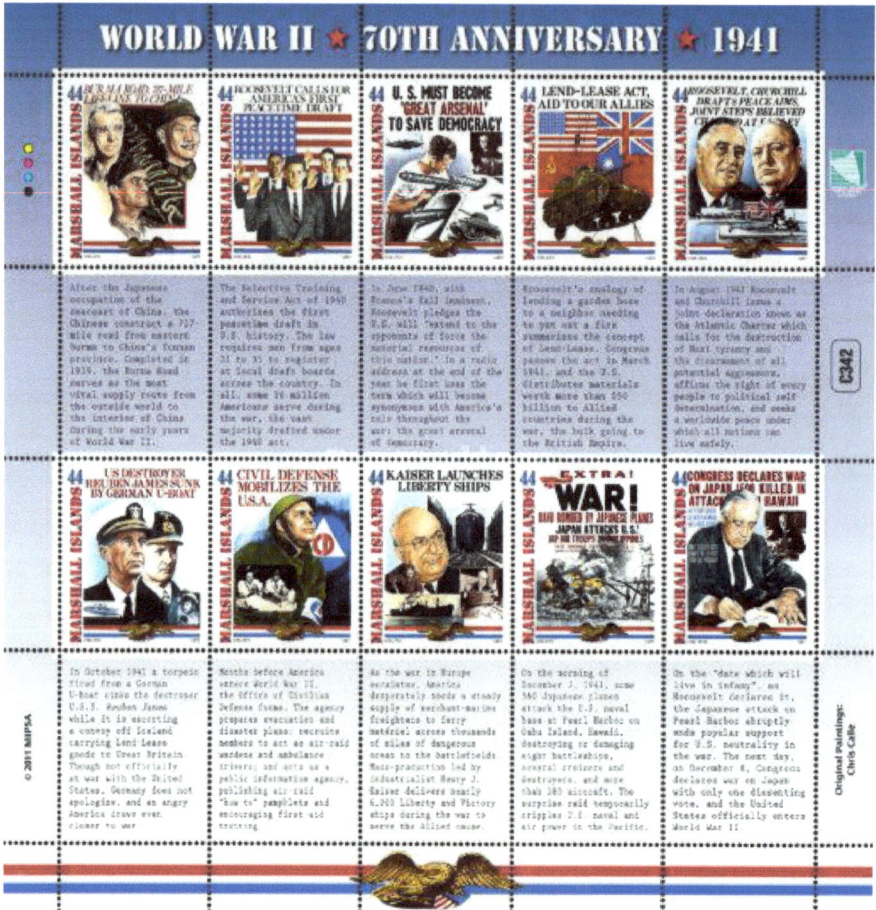

C342 : Marshall Islands 2011 : The 70th Anniversary of the US Joining World War II.

Stylised newspaper headings and pictures :

          Burma Road link to China

          Roosevelt calls for America's first draft

          US must become 'great arsenal to save democracy

          Lend-Lease Act, Aid to our Allies

          Roosevelt and Churchill draft peace plans

          Civil Defence mobilises USA

          Kaiser launches Liberty Ships

          WAR! Japan attacks USA

          Congress declares war on Japan.

C351 :Marshall Islands 2012 : The 70th Anniversary of World War II- 1942.

Stylised newspaper headings and pictures :

Tokyo bombed

Roosevelt praises U S rationing effort

US wins great battle in the Coral Sea

US warships defeat Japan

Code-breaking turning the tide in the Pacific

Japanese fleet smashed by US at Midway

Women in the War - Maisie the riveter

Marines gain hold on Solomon Islands

Armed Forces land in French Africa.

C370 : Marshall Islands 2013 : The 70th Anniversary of World War II – 1943.

Stylised newspaper headings and pictures :

        Tenth Fleet : Overcoming the U-boat menace

        US Medical Units saving lives on front line

        Allies invade Sicily : Patten's 7th Army drives on Palermo

        B24 Liberators bomb prestigious oilfields

        V-Mail : Letters from around the globe

        Allied Forces land at Salerno

        War Loans

        Willy and Joe cartoons amuse.

        Gold Star – plan for freedom

        US Marines land at Tarawa Atoll.

C370 : Marshall Islands 2014 : The 70th Anniversary of World War II - 1944.

Stylised newspaper headings and pictures :

> Allied Forces take New Guinea
> Allied bombing of Germany
> "Invasion is on"
> Marine Units in ground attacks
> Submarine warfare cripples Japan's shipping routes
> Paris and Rome liberated
> US Navy, Admiral Halsey featured
> *Red-Ball Express* speeds supplies
> Battle for Leyte Gulf
> Battle of the Bulge.

Marshall Islands 2015 : The 70th Anniversary of World War I - 1945.

Stylised newspaper headings and pictures :

US Marines steam ahead

MacArthur back in Luzon

Marines land in Okinawa

US and Soviet soldiers link at River Elbe

Allied Forces liberate concentration camps

The War in Europe is ended

The War uproots millions

First atomic bomb dropped on Japan

WAR OVER

The War is ended.

The other contender for third best thematic set.

United States 1991 : "A world at war". Miniature sheet.

This image is the first of a series of five miniature sheets, each comprising a map and ten fiscally valid service fees – within US territories.

Atlantic Charter sets war aims of allies

America becomes "arsenal of democracy"

Destroyer Reuben James sunk October 31

Civil Defense mobilizes Americans at home

First Liberty ship delivered December 30

Japanese bomb Pearl Harbor, December 7

U.S. declares war on Japan, December 8

United States 1991 : "A world at war"/ 1941. The 10-stamp set.

I have shown the ten images separately to reflect their actual size and make it easier to read the text – the context of the image within the story being told in this, the first of five historical sets illustrating World War Two.

The maps, central to showing the context of the issue illustrate the theatres of activity and the movement of US forces in the conflict for the year being considered. The text on the maps has not scanned well.

The other four issues within this history are entitled :

United States 1992 : "Into the Battle / 1942.
United States 1993 : "Turning the Tide" / 1943
United States 1994 : "Road to Victory" / 1944
United States 1995 : "Victory at Last" / 1995.

Another possible contender as a best set - another Fleetwood product

Marshall Islands 2016 : Honouring the Art of Howard Koslow, 1924-2016.

Marshall Islands 2016 : The 75th Anniversary of the Pearl Harbor Attack.

A book with almost the same title as the Fleetwood #1 set of images

Interestingly the British author Peter Darmon has published in 2016 his own determination of "100 Events that shaped World War II. The two covers within which the 224 pages of the book was published are shown below :

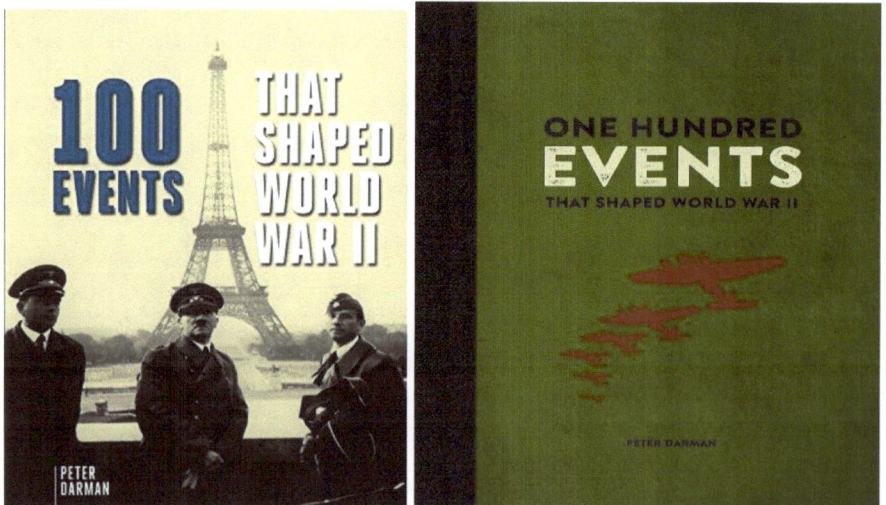

"A compendium of the 100 most decisive events, both on and off the battlefield, that shaped the course of the conflict between 1939 and 1945. The bloodiest conflict in human history was not decided by the actions of an enlightened few. Instead, it played out as countless acts of heroism, on battlefields across the planet, in events whose outcomes were anything but certain. Within this history of individual sacrifice, it is possible to identify landmark events that shifted the momentum of World War II, from the early gains of Hitler's Axis powers to the growing Allied might of the later years...One Hundred Events That Shaped World War II tells the story of the war. Taking into account social, economic, technological and political factors, it presents the people involved and identifies the strategic masterstrokes, as well as blunders, that determined the ultimate course of victory".

How consistent are the selection of the 100 epic — war-shaping events? 1995 and 2016 examples

Year	Fleetwood stamp image	Darman / event shaping
1939	Invasion of Poland	Nazi-soviet Non-Aggression Pact
	Sinking of HMS Royal Oak	The Red Orchestra
	Invasion of Finland	Britain and France declare War
	Battle of the River Plate	Britain establishes the MOI
		The Winter War
		The Phony War

1940	Invasion of Denmark and Norway	Rationing begins
	Katyn Forest massacre	The *Altmark* incident
	Churchill becomes Prime Minister	Churchill becomes Prime Minister
	Low Countries invaded	The Battle of Sedan
	Deliverance at Dunkirk	Hitler halts the Panzers
	Occupation of Paris	The miracle of Dunkirk
	Battle of Mers-el-Kebir	Seven Days in June
	Burma Road conflicts	Cracking the German Military Codes
	UK/US destroyers for bases Pact	Britain introduces a Radar Network
	The Battle of Britain	The Battle of Britain
	Tripartite Pact signed	Convoy HX-72
	Roosevelt re-elected to third term	Britain introduces Sonar
	Battle of Taranto	Italian fiasco in Greece
		Operation Judgement

1941	Roosevelt's Four Freedoms speech	US Congress / Lend-Lease Act
	Battle of Beda Fomm	Enter the Desert Fox
	Germany invades Balkans	Balkan Interlude
	Sinking of the Bismark	Soviet-Japanese Neutrality Pact
	German invasion of Russia	The Capture of U110
	Atlantic Charter	Germany invades Russia
	Siege of Moscow	Moving Soviet Industry East
	USS Ship Rueben James sunk	The Battle of Moscow
	Japanese attack Pearl Harbour	The Defence of Leningrad
	Japanese capture Guam	Women on the Home Front
	Fall of Singapore	Japan brutally represses the Chinese
	First combat by "Flying Tigers"	The attack on Pearl Harbour
	Fall of Wake Island	Japan conquers Malaysia and Singapore
		The Philippines are lost
		Germany declares Total War
		Saving the Suez Canal

Year	Fleetwood stamp image	Darman / event shaping
1942	Arcadia Conference	The Wannsee
	Fall of Manila to Japanese Forces	Area Bombing Directive
	Japanese take Rabaul	Heavy Bombers enter the fray
	Battle of Java Sea	Fall of Burma
	Rangoon falls to Japanese	Bataan Death March
	MacArthur evacuates Corregidor	War Bonds are introduced
	Raid on Saint Nazaire	The Doolittle raid
	Surrender of Bataan/Death March	US gets warning of the Midway attack
	Doolittle raid on Tokyo	The Battle of the Coral Sea
	Fall of Corregidor	Assassination of Reinhard Heydrich
	Battle of the Coral Sea	The Battle of Midway
	Battle of Midway	The OSS is established

	Village of Lidice destroyed	The fall of Tobruk
	Sevastopol falls to Germans	The Battle of Guadalcanal
	Convoy PQ-17 destroyed	The Battle of El Alamein
	Marines land on Guadalcanal	Operation Torch
	Battle of Savo Island	Operation Uranus.
	Dieppe Raid	Scuttling the French Fleet
	Battle of Stalingrad	Japan's Oil Dilemma
	Battle of the Eastern Solomons	The Atomic Age begins
	Battle of Cape Esperance	Battle of the Barents Sea
	Battle of El Alamein	
	Battle of Barents Sea	

1943	The Casablanca Conference	The Casablanca Conference
	Liberation of Khakov	Heavy Water sabotage
	Battle of the Bismark Sea	Battle of Karkov
	Interception of Yamamoto	The Warsaw Ghetto Uprising
	Battle of Kursk	Operation Mincemeat
	The Allied invasion of Sicily	Axis collapse in North Africa
	Bomber raids on Schweinfurt	The Trident Conference
	Liberation of Smolensk	The Allies strike a blow in the Atlantic
	Landing at Bougainville	Russian Partisans defy the Axis Forces
	US Forces invade Tarawa	The Pointblank Directive
	Teheran Conference	The Battle of Kursk
	Battle of North Cape	The invasion of Sicily
		Mussolini overthrown
		The Invasion of Italy
		US Forces land on Gilbert Islands
		Long-Range fighter escorts introduced

1944	SHAEF Commander named	Operation Carpetbagger
	Invasion of Anzio	Gustav Line is breached
	Siege of Leningrad lifted	Operation Flintlock
	US liberates Marshall Islands	The Battle for Kohima and Imphal
	Japanese defeated at Truk	Operation Ichigo
	"Big Week"- US bombs Germany	US Forces land on Ohama Beach

1944	continued	
	Rome falls to the Allies	Air War over Normandy
	D-Day and Normandy landings	The Battle of the Philippine Sea
	V-1 bombardment of England	Operation Bagration
	US Marines land on Saipan	The Arsenal of Democracy emerges
	First Battle of the Philippine Sea	The July Bomb Plot
	US liberates Guam	The Blockade of Japan
	The Warsaw Uprising	The desertion of Hitler's Axis Allies
	Liberation of Paris	The Warsaw Uprising
	US marines land on Peleliu	Operation Market Garden
	MacArthur returns to Philipinnes	The Leyte Landings
	Battle of Leyte Gulf	The Battle of Leyte Gulf
	German Battleship Tirpitz sunk	Operation 'Watch on the Rhine'
	Battle of the Bulge	

1945	Yalta Conference begins	The Yalta Conference
	Bombing of Dresden	The Battle of Iwo Jima
	Iwo Jima invaded by US Marines	Fire raids on Japan
	Remagen Bridge taken by US	The Battle of Okinawa
	Okinawa invaded by US Forces	The Battle of Berlin
	Death of Franklin D Roosevelt	Italy surrenders
	US/USSR Troops at Elbe River	Hiroshima
	Soviet troops conquer Berlin	Operation August Storm
	Allies free Concentration Camps	
	V-E Day	
	United Nations Charter signed	
	Potsdam Conference convenes	
	PM Churchill resigns	
	Atomic Bomb on Hiroshima	
	V-J Day	

The two 100 event lists compared. The same stories on both lines have a blue background.

I had expected a more than a 26% match within the two 'histories'. The Fleetwood Military Historical Panel, in 1995, have been aware that they were consolidating images into list of 100 epic events. The English-based former research officer working independently, has sought to provide a broader perspective in shaping his interpretation of World War II history. A bigger picture will change the future 'shape' of infrastructure and weapons as technology develops.

The Fleetwood military historians were conscious of the leadership issues and recognise the principal players and their contributions.

Commemorative postage stamps as I have studied them here – and as may have been described by Henry Reynolds, an early Australian historian as - *carnivals of remembrance.* The images may be vivid but they are static and etched into a static world. Today's historian, such as Peter Darman, can view the bigger picture. As Peter Stanley has reminded us within his 2024 work, "Beyond the Broken Years – Australian military history in 1,000 books" there are many people and publishers wanting to share their ideas. We know that perspectives will change over time. That is what makes history so persuasive.

I stick to postage stamps. They may grow old, but at the time of issue, they reflected what a public body, in most cases, believed to be the real story. That the best set ever was written and published by a commercial endeavour on behalf of a small legislative, newly created, nation makes it even more remarkable.

References.

The One hundred Epic Events of World War II in stamps (1995), Cheyenne, Wyoming, USA, Fleetwood Publications. ISBN 0-91337650 -7.

Black, Jeremy (2020), A History of the Second World War in 100 Maps. University of Chicago Press. ISBN 978-0-22675524-3.

Chappell, John David (1997). *Before the bomb: how America approached the end of the Pacific War.* University of Kentucky Press. ISBN 978-0-8131-1987-8.

Darman, Peter (2016) The 100 Events that shaped WW2 (2016), UK, Apple Press.
ISBN 978-1- 8454364-4-5.

Journal of the American First Day Cover Society (2004). Volume 49, Number5 dated July 15, 2004. Majander, Jari (2020), In Quest for abusive and undesirable stamp issues. 2020 TC News (www.fipthematicphilately.org/2020/TCNewsNo29September2020.pdf).

Perry, Roland (2024), The Battle of the Generals – MacArthur, Blamey and the Defence of Australia in World War II. Crows Nest, NSW. ISBN 978-1-76147-049-3.

Stanley, Peter (2004), Beyond the Broken Years. Australian Military History in 1,000 books. Sydney NewSouth Publishing. ISBN 978-1- 7611701-4-0.

Yardley, Christopher B (2019), A Great War Study – The Centenary commemorative postage stamps 2014-2018, Canberra, Cannava House Publications / Ingram Spark. ISBN 978-0-6486-6710-0.

Yardley, Christopher B (2022), The Second World War – representing world conflict on postage stamps, Canberra, Cannava House Publications / Balboa Press. Volume One ISBN 978-1-9822-9297-3. Volume Two ISBN 978-1-9822-9299-7.

On my reading list : Konstam, Angus (2015). *Taranto 1940; The Fleet Air Arm's precursor to Pearl Harbor.* Osprey Campaign Series #288. Osprey Publishing. ISBN 978-1-4728-0896-7.

Acknowledgements.

Having come up with the idea of looking at my favourite set / series of stamps EVER I approached Wendy Buckle, editor of *Themescene*, the journal of the British Thematic Society to ask how she might go about establishing the viability of the idea.

Wendy suggested I consult the expert panels of the UK Royal Philatelic Society and / or the Smithsonian Institute in Washington. Both panels responded positively. Daniel A. Piazza, Chief Curator of Philately at the Smithsonian National Postal Museum gave me a contact address for James A Willms, a former CEO of Unicover Corporation, the publisher of the series.

In turn, Mr. Willms put me in touch with John Clark Helzer. He responded, "I am thrilled to hear from you! Thank you so much for writing to me, and for your interest in my late father James A. Helzer's brainchild *The One Hundred Epic Events of World War II In Stamps* postage stamp series issued by the Republic of the Marshall Islands (RMI), which was conceived of, designed, created and printed by my father's company Unicover Corporation (under the trademark of its flagship Fleetwood division) in the early years of its three-decade role as the Stamp Issuing Bureau of the Marshall Islands Postal Service Administration".

John had worked with his father, who had sadly passed away in 2008. John had worked in the role as a Project Manager for the series issue.

I had long thought the set was so significant it would have had a separate "history" prepared. It had. John had a copy of the commemorative history of the series which he sent to me, and which features in my text.

Seeking to build a 'virtual' set of stamp images I have extensively copied images from "Stampworld" and the vendors of Marshall Island stamps on eBay. eBay vendors have been particularly helpful particularly 'fishmull' and 'Historiaphil' have entered a discussion with me. I have used the expertise of specialist Pacific stamp dealer, (Lisa), of the Brookman Stamp Company of Vancouver, WA, as my guide to the cataloguing and Scott numbering system.